Internet of Things

Technology, Communications and Computing

Series Editors

Giancarlo Fortino, Rende (CS), Italy

Antonio Liotta, School of Computing, Edinburgh Napier University, Edinburgh, UK

The series Internet of Things - Technologies, Communications and Computing publishes new developments and advances in the various areas of the different facets of the Internet of Things. The intent is to cover technology (smart devices, wireless sensors, systems), communications (networks and protocols) and computing (theory, middleware and applications) of the Internet of Things, as embedded in the fields of engineering, computer science, life sciences, as well as the methodologies behind them. The series contains monographs, lecture notes and edited volumes in the Internet of Things research and development area, spanning the areas of wireless sensor networks, autonomic networking, network protocol, agent-based computing, artificial intelligence, self organizing systems, multi-sensor data fusion, smart objects, and hybrid intelligent systems.

Indexing: *Internet of Things* is covered by Scopus and Ei-Compendex **

Franco Zambonelli • Giancarlo Fortino •
Barbara Re • Mirko Viroli
Editors

Fluidware

Novel Approaches for Large-Scale IoT Systems

Springer

Editors
Franco Zambonelli
Università di Modena and Reggio Emilia
Reggio Emilia, Italy

Giancarlo Fortino
Università della Calabria
Rende (CS), Italy

Barbara Re
Università di Camerino
Camerino, Italy

Mirko Viroli
Alma Mater Studiorum - Università di Bologna
Cesena, Italy

ISSN 2199-1073 ISSN 2199-1081 (electronic)
Internet of Things
ISBN 978-3-031-62145-1 ISBN 978-3-031-62146-8 (eBook)
https://doi.org/10.1007/978-3-031-62146-8

This Springer imprint is published by the registered company Springer Nature Switzerland AG
The registered company address is: Gewerbestrasse 11, 6330 Cham, Switzerland

Here's to all the researchers that have participated to the activities of the Fluidware project, thanks for four exciting years of collaborative research.

No PhD students were harmed during the making of this book.

Preface

In the past few years, progress in the area of wireless communications, sensors, actuators, and microcontrollers has made it possible to embed computing and communication systems in physical objects and environments, with the capability of sensing and affecting in real-time the physical and social world. In other words, the vision of the "Internet of Things" (IoT) has rapidly become an everyday reality and a source for many application areas.

However, IoT applications and services can be hard to program and deploy. Indeed, services and applications for the IoT may involve composing and coordinating myriads of heterogeneous devices, typically dispersed over an environment. This introduces many engineering challenges, such as the need to support scalability, very frequent events and changes, faults of devices, and the continuous evolution of the scenario due to novel devices being deployed and others being dismissed.

In the above context, the IoT research community worldwide started putting notable investigation efforts toward the identification of novel programming approaches for IoT applications that could make it possible to deal with the above sources of complexity automatically and rather enable programmers to easily program and deploy flexible and adaptive applications, enabling to express in full the potentials of the IoT fabric.

This book is the direct result of one such investigation effort. This book aims to collect the key research results obtained during the 3 years of Italian "Fluidware" research project activity.

The Fluidware project is one of the projects funded by the Italian Ministry of University and Research (MUR) with the framework of the PRIN (Programmi di Ricerca di Interesse Nazionale) 2017 research program. Fluidware was selected after a very competitive call for proposals and was ranked second out of more than 200 proposals submitted. The project started in September 2019 and was initially planned for 2 years. However, the COVID pandemic and the many lockdowns slowed down research worldwide, Italy included. The ministry decided to extend PRIN 2017 projects until August 2023, making the project duration 4 years.

The Fluidware consortium included four research units from different Italian universities:

- The University of Modena and Reggio Emilia, and in particular the Pervasive Intelligence research group led by Franco Zambonelli, who was also the coordinator of the whole project
- The University of Bologna, and in particular the research group on aggregate computing led by Mirko Viroli
- The University of Calabria, and in particular the research group on IoT led by Giancarlo Fortino
- The University of Camerino, particularly the PROS—PROcesses & Services lab of the Computer Science Division led by Barbara Re

The specific names of all the researchers who participated in the project and contributed results to it include all the authors of the chapters that constitute this book. Such chapters are intended to comprehensively overview the different research topics and research results of the Fluidware project and is logically organized into three parts:

- Part I introduces the general ideas behind Fluidware and an overview state of the art in the area of programming models for the IoT. In particular: Chap. 1 presents the vision and overall ideas behind Fluidware, as well as a case study in the area of assisted living; Chap. 2 surveys the state of the art of IoT programming, trying to clarify which programming paradigms and platforms are most commonly adopted, with the goal of uncovering which research areas are mostly active in IoT programming.
- Part II presents the different technologies and methods developed throughout the project. In particular: Chap. 3 describes the middleware architecture developed for supporting Fluidware computing via the pulverization paradigm; Chap. 4 introduces a software engineering methodology to support the development of Fluidware applications and, more in general, IoT applications; Chap. 5 describes how business process management technologies can effectively act as a powerful tool for model-driven engineering of Fluidware systems; Chap. 6 presents the distributed algorithms based on aggregate computing that can be used to support adaptivity of Fluidware application along the space and the time dimensions.
- Part III overviews the many research challenges and visions that have emerged during the execution of the project and that represent potentially fruitful areas for future research. In particular: Chap. 7 shows how Fluidware concepts can be exploited to effectively develop IoT systems in the emerging scenario of digital twins; Chap. 8 introduces a novel development toolchain that, in the future, can support the formal development of Fluidware systems; Chap. 9 presents a potential extension of Fluidware concepts to support more adaptable and open

IoT systems; finally, Chap. 10 overviews the potentials of learning technologies to develop Fluidware applications capable of dynamically self-adapting to the context.

Hope you will enjoy the reading!

Rende, Italy — Giancarlo Fortino
Camerino, Italy — Barbara Re
Cesena, Italy — Mirko Viroli
Reggio Emilia, Italy — Franco Zambonelli
January 2024

Acknowledgments

We thank the Italian MUR for having supported the Fluidware activities in the context of the PRIN 2017 research initiative.

Contents

Contributors

Gianluca Aguzzi DISI, Alma Mater Studiorum—Università di Bologna Cesena, Italy

Roberto Casadei DISI, Alma Mater Studiorum—Università di Bologna Cesena, Italy

Flavio Corradini Computer Science Division, University of Camerino Camerino, MC, Italy

Nicolas Farabegoli DISI, Alma Mater Studiorum—Università di Bologna Cesena, Italy

Fabrizio Fornari Computer Science Division, University of Camerino Camerino, MC, Italy

Giancarlo Fortino Department of Computer Science, Modeling, Electronics and Systems Engineering (DIMES) of the University of Calabria Rende, CS, Italy

Stefano Mariani Dipartimento di Scienze e Metodi dell'Ingegneria, Università di Modena e Reggio Emilia Reggio Emilia, RE, Italy

Sara Pettinari Computer Science Division, University of Camerino Camerino, MC, Italy

Danilo Pianini DISI, Alma Mater Studiorum—Università di Bologna Cesena, Italy

Barbara Re Computer Science Division, University of Camerino Camerino, MC, Italy

Lorenzo Rossi Computer Science Division, University of Camerino Camerino, MC, Italy

Claudio Savaglio Department of Computer Science, Modeling, Electronics and Systems Engineering (DIMES) of the University of Calabria Rende, CS, Italy

Francesco Tiezzi Dipartimento di Statistica, Informatica, Applicazioni, University of Florence Florence, Italy

Mirko Viroli DISI, Alma Mater Studiorum—Università di Bologna Cesena, FC, Italy

Franco Zambonelli Dipartimento di Scienze e Metodi dell'Ingegneria, Università di Modena e Reggio Emilia Reggio Emilia, RE, Italy

Part I
Overview

Chapter 1
Fluidware: An Approach Toward Adaptive and Scalable IoT Systems

Giancarlo Fortino, Barbara Re, Mirko Viroli, and Franco Zambonelli

1.1 Introduction

This chapter originally appeared in 2019 in [29], at the time the Fluidware project was going to start and represents a sort of "manifesto" for Fluidware itself.

In Fluidware, we consider not so far future "Internet of Things" (IoT) environments, in which our everyday objects and our domestic and urban environments will be densely enriched with sensing, computational, and actuating capabilities [5, 33, 34]. Such "IoT fabric" can potentially become a very powerful infrastructure that, if properly programmed, can provide a variety of smart IoT services and IoT applications (intended as coherent suites of services): They can work both at the personal and at the community level and can help us better interact with the cyber-physical world and the objects in it. However, unveiling the true potentials of large-scale IoT deployments requires novel and dedicated models and programming abstractions [9, 40, 51].

G. Fortino
Università della Calabria, Rende, CS, Italy
e-mail: giancarlo.fortino@unical.it

B. Re
Università di Camerino, Camerino, MC, Italy
e-mail: barbara.re@unicam.it

M. Viroli
Università di Bologna, Cesena, FC, Italy
e-mail: mirko.viroli@unibo.it

F. Zambonelli (✉)
Università di Modena e Reggio Emilia, Reggio Emilia, RE, Italy
e-mail: franco.zambonelli@unimore.it

F. Zambonelli et al. (eds.), *Fluidware*, Internet of Things,
https://doi.org/10.1007/978-3-031-62146-8_1

Programming services and applications for the IoT fabric may involve the composition and coordination of a multitude of heterogeneous devices, possibly dispersed over a wide and/or articulated area—e.g., from smart houses, buildings, and hospitals to whole factories or even smart cities. However, this scenario embeds many sources of complexity: the potential high number and density of deployed devices, the high frequency of computational events to occur and be managed, the need to self-adapt to short-/medium-/long-term changes and faults, the need to operate on top of a dynamically evolving infrastructure comprising IoT/edge/cloud devices and resources, the need of promptly (often in quasi-real time) reacting to sophisticated space–time situations, and the potential of running complex, goal-oriented orchestrations of distributed activities across devices of heterogeneous computational power (from tiny devices to cloud servers).

The above issues can make it very hard to develop and deploy dependable services and applications exhibiting predictable behavior, especially when using traditional service composition approaches that would require explicit handling of a multitude of possible exceptions and alternate scenarios. Accordingly, there is compulsory need of novel programming approaches that make it possible to abstract from mundane complexity details and rather rely on high-level constructs enabling to express the potentials of the IoT fabric in full.

To address such issues, we envision a novel architectural and programming model for IoT services and applications (along with the associated supporting platform, engineering methodology, and tools) conceived to ease the development of flexible and robust large-scale IoT services and applications. Starting from previous findings in the areas of attribute-based coordination [1, 2, 22], field-based coordination [32, 47], collective adaptive systems [4, 41], and stream computing and aggregate computing [7, 9], the proposed "Fluidware" framework address the complexity of building modern, large-scale IoT systems, by a full-fledged engineering approach revolving around a new notion of distributed programming.

Our approach relies on the idea to abstract collectives of devices of the IoT fabric (sensors, actuators, and other edge/cloud devices) as sources, digesters, and targets of distributed flows of contextualized events, carrying information about data produced and manipulated over time. Accordingly, programming services and applications implies specifying "funnel processes" to channel, elaborate, and redirect such flows in a fully distributed way, as a means to coordinate the activities of devices and realize services and applications.

Funnel processes can be specified in a declarative way, in terms of how they consume and produce events over space and time. They can be associated with distributed and contextualized streams (i.e., flows) of events in terms of advanced pattern matching and filtering mechanisms based on semantics of data and space–time conditions on their production. Thus, the specification of funnel processes totally abstracts from the actual devices to which events belong: It can be such to include not only a limited number of local streams from a limited number of devices but also a large number of devices spread over a large scale, enabling to define scale-independent computational activities inherently addressing self-adaptation to contextual conditions and smoothly fitting various computing/network infrastructures. Indeed, a proper supporting platform/middleware [37] will need to be put in place to take care of the actual deployment of funnel processes, of their allocation at the level of IoT devices and/or edge/cloud computers [42], of their optimized execution, and of their transparent replication and distribution to opportunistically cover large-scale areas.

Overall, the Fluidware approach, and the tools and techniques developed in its context, can support the bottom-up construction of complex IoT applications through a correctness-guaranteed stack of software components, incrementally encompassing basic building blocks of stream/event manipulation, libraries for distributed coordination, and reusable IoT services.

In order to present the key ideas of the Fluidware approach, the remainder of this chapter is organized as follows: Sect. 1.2 defines an application context on ambient assisted living, Sect. 1.3 consequently derives challenges for novel programming approaches, Sect. 1.4 sketches the Fluidware architectural and programming model and how it would address those challenges, Sect. 1.5 defines the path toward the full development of Fluidware and the research activities required, Sect. 1.6 discusses related work in the area, and finally Sect. 1.7 summarizes contributions of the work and ongoing developments.

1.2 A Case Study Scenario

As a reference to ground discussion, we consider an Ambient Assisted Living (AAL) scenario which we call the "Fluidware house," where the full potential of the IoT is exploited to improve Mark's daily life.

> Mark is a 65 years old retired person, married with no children. He had an ischemic stroke with paralysis of the right upper and lower limbs 14 months ago. Mark was discharged from the rehabilitation center 3 months after the stroke, and then he continued to rehabilitate at home. Once his socio-health situation was evaluated, the psychiatrist and social services have suggested Mark and his wife to move to a Fluidware house, more suited to the needs of the family.

In a Fluidware house, we assume that all the rooms are densely enriched with connected sensors and actuators: light and heat controllers, gas and smoke detectors, presence and motion sensors, pressure sensors located on the sittings, electric energy consume sensors, and so on. While the environment is monitored, the person is constantly controlled by wearable sensors that are able to detect vital signs (e.g., heart rate and respiration, activity level, and posture). In such a scenario, IoT devices can be exploited to realize a variety of different services to support care activities of individuals, to help individuals and their family members in their everyday self-managed health-care activities, and to control the overall conditions of the house.

> Mark's day begins at 8 am. After waking up, he is able to get out of bed with the help of his wife. Mark is able to walk short distances with orthosis and a stick support and can reach quite easily the bathroom where he can wash in a bathtub with a rising platform, always with the help of a caregiver. The morning is the part of the day dedicated to rehabilitation

aimed primarily at preventing complications due to hypomobility of limbs and to increase physical performance. During this phase, as for the rest of the day, it is very important to measure vital parameters through a wearable system of sensors.

In this situation, the Fluidware house will use sensors to monitor the activities carried out, which are then involved in the assessment of the condition of normality/abnormalities. The normal condition implies that at least one sensor registers the presence of the person through the recognition of a movement. Terms of inactivity (e.g., TV watching, sleeping) are considered normal if they involve the use of furniture such as chairs or bed with a pressure sensor that detects the use. Abnormality conditions can depend on prolonged inactivity or on alarms related to wearable sensor data exceeding thresholds.

The afternoon is the time for relax and the presence of a caregiver within the home is not essential.

Here, the environmental sensors (e.g., luminosity, temperature, humidity, and gas) monitor the conditions inside the house. The supplied data are used by Fluidware service to automatically adjust the environment conditions, optimizing the use of energy in the rooms where Mark will not typically move to.

After lunch, Mark's wife leaves her husband alone in the house. Mark gets up from his chair and with the help of the stick goes into the bathroom to wash his hands. The floor is slippery and he falls. Having mobility problems, he is no longer able to stand up alone.

In the unpleasant situation of a fall, we can have the intervention of several sensors. The wearable sensors in place are able to detect the posture (vertical or horizontal) and the acceleration along three axes [25]. This is particularly useful to determine emergency conditions such as the occurrence of a fall [26]. Present sensors are able to both detect walking movements in the selected detection zone and respond to them accordingly. Moreover, motion sensors respond to the tiniest movements using extremely high resolution and precision sensors detecting the movement of arms and fingers. The condition of horizontal position and the detection of a downward acceleration are considered fundamental situations because they are considered an abnormality. If, over a short period of time, the sensors do not detect a new acceleration upward and if the person remains in a horizontal position without moving, the Fluidware emergency service can say, with a certain degree of certainty, that the person has fallen. The overall coordination of emergency service involving the different caregivers can also be implemented exploiting Fluidware potentialities.

Each day, Mark's doctor receives summary information about Mark's activity and vital parameters both on its PC and its smartphone. She can check them, compare them with those of previous days or weeks to analyze trends, and can then even command changes in the configuration of the house or of the care program.

All the information provided by the several sensors deployed in the house gets automatically aggregated, filtered, and channeled to the cloud. Moreover, in the discussed scenario considering a lot of information exchange taking place, it is worth to notice the need to address privacy/security issues. Additional processes living in the cloud take care of redirecting general information and events to Mark's doctor.

1.3 Challenges

The key motivations for Fluidware are to address a number of issues that currently hinder the possibility to easily develop complex and large-scale IoT systems and applications. The case study scenario we considered helps us identify the following challenges:

Device Independence Current IoT approaches are highly device-dependent, assuming the existence of specific types of devices in specific locations. However, the heterogeneity of devices, their ephemerality, and the impossibility to individually control all devices in an environment require programming approaches enabling services to be programmed and expressed in device-independent terms.

In fact, Mark's house seamlessly manages the coexistence of different types of devices (e.g., motion, environment, and medical sensors) in different environments and with heterogeneous characteristics (e.g., kitchen, bedroom, and bathroom introduce their own space constraints).

Scalability IoT services may not only involve individual sensors and actuators or compose and coordinate a limited number of local devices but also exploit a multitude of cooperating devices distributed over a wide area. An approach for programming and deploying IoT services should adopt the same basic model for small- and for large-scale services and should not lose in effectiveness when scaling.

As Mark's house is densely enriched with connected IoT devices, cooperating to provide services, scalability is required for a number of reasons: Mark's personal condition can quickly change, asking for prompt change also in the environment as an increasing need of monitoring Mark's daily routine; novel IoT devices can be added into the house to provide applications that impose additional traffic overhead. As a result, there is need to handle a growing amount of data in order to automatically manage high-quality information on health and environment status.

Adaptivity IoT systems are called to operate in very dynamic environments. Devices can be faulty, mobile, or become unreachable. Yet, IoT services and applications must adaptively react and be highly available to serve reliably despite contingencies.

The house is a live dynamic environment, where devices change their sensing and actuating capability as a result of the changing condition of Mark's personal status. Different types of change need to be supported: There are situations not known a priori that are partially repeatable, unpredictable, and emergent, while others are more structured, repeatable, and predictable; overall, there is need to be able to support both a self-adaptive management of daily routines and adaptation in the sense of human-controlled (re)configuration.

Seamless Integration of Devices, Edge, and Cloud Levels The overall IoT fabric will include a multitude of distributed devices, edge computers that can act as "local clouds" for the devices in a locality, and general cloud resources. To effectively exploit the IoT fabric, there is thus need of supporting direct device-to-device interactions, edge computation, and cloud resources, all within the same design and programming abstraction.

Effectively monitoring and controlling Mark's house means supporting different types of integration: Direct device-to-device interactions will be needed in order to implement local alarms (e.g., exceeding a temperature threshold), edge computation to locally enact complex situation recognition (e.g., fall detection), and cloud resources to store historical information and interact with external stakeholders (e.g., caregivers and doctors).

Interoperability and Security Enabling interoperability among heterogeneous devices and ensuring security of IoT systems and services are key challenges to enable the widespread diffusion of IoT services.

Interoperability [24] needs to be guaranteed in order to share information between devices in the house (e.g., the daily routine involves several objects dedicated to understand normal and emergency situation), as well as between the house and external environment (e.g., to interact with the hospital in case of emergency), which requires to consider the multitude of standards applicable to assistive technologies and the security and privacy issues concerning the personal nature of the scenario.

1.4 The Fluidware Approach

The Fluidware approach considers IoT-enriched environments densely populated with a variety of IoT devices (overall forming the "IoT fabric"), acting as sensors or actuators or both. Sensing devices generate "contextualized streams" of data representing events about something that is happening somewhere in the environment. Actuating devices receive "contextualized command streams" related to how they should affect the environment, and people within, over time.

1.4.1 Funnel Processes

The key idea of Fluidware is that IoT services and applications can be realized by transformations of widely distributed streams of events, which we also call "event flows" (or simply "flows"), involving collectives of devices. This is achieved by means of an abstraction of "funnel process," acting as digester and producer of "event flows," elaborating/sending/(re)distributing them over the network of distributed IoT devices or over the edge/cloud (see Fig. 1.1).

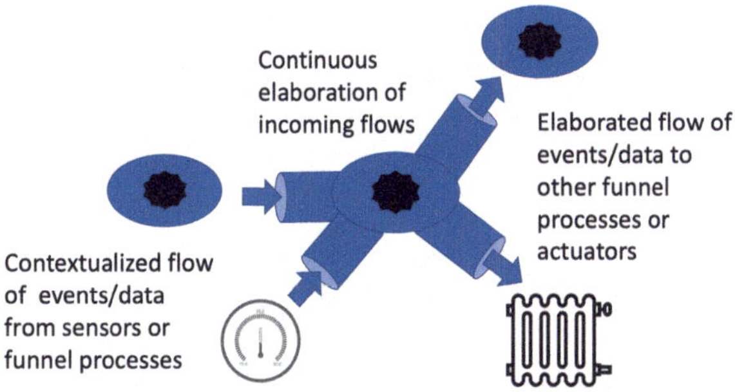

Fig. 1.1 A general abstract representation of funnel processes exploited locally

Funnel processes will be specified in a declarative way independently of their actual allocation and distribution (managed by the Fluidware middleware). Their specification will be such to define which flow to connect to, depending on contextual/spatio-temporal and semantic matching (e.g., from "in room X now" to "where average temperature has been greater than 25 Celsius degrees in the last 5 minutes"). Thus, a funnel specification will totally abstract from the actual devices that are sources, manipulators, or targets for events: It can be such to include not only a limited number of local events from a limited number of devices but also a large number of devices spread over a large scale. This enables to define scale-independent computational activities, inherently independent of external conditions and smoothly fitting various computing/network infrastructures. Most specifically, an event flow is manipulated computationally to produce a new flow, by a combination of mechanisms typical of stream processing frameworks and of self-adaptive and self-organizing systems, such as fully distributed and resilient aggregation, spreading, persistence, and so on [46].

In particular, we envision the definition of a library of primitive funnel templates, including:

- *Filters* that select only the events that match certain criteria, expressed via conditions on attributes concerning event source, location, time, content, and so on [22];
- *Aggregators* that cluster events into groups (based on correlations on their attributes) and correspondingly create new events for each group (creating average values over time or space, reunifying complex multi-event situations into alert events, and so on)
- *Transmuters* that manipulate the content of events leaving their overall structure unchanged (resetting content, applying increments, mapping, and so on)
- *Relocators* that move source/location/time of events (to specify space–time relocation policies for event streams)

Then, special operators and combinators will be used to compose and interconnect funnel processes and attach them to physical elements of various kinds, such as:

- *Splitters* that take one stream and divide it in multiple streams

- *Mergers* that conversely combine streams
- *Pipes* that sequentially compose funnels
- *Hooks* that connect funnels to virtualized devices (macro sensors/actuators/computational resources [11])

1.4.2 Deployment Scenarios

At the implementation and deployment level, the Fluidware platform/middleware will take care of creating actual distributed processes and connecting them to the actual flows of events and the physical devices supporting them, to realize funnel process specifications. Such processes can be deployed on any IoT device with enough resources to support their execution. Opportunistically, funnel processes can be deployed at the level of edge computers associated with some specific location and having access to all devices in that location [40, 44] or even at the level of some centralized cloud.

Clearly, when allocated to individual devices, a process can predigest and elaborate the data sensed by that device before forwarding it to some other devices or to some collection point. However, it can also be seamlessly used to spread and redistribute events to nearby devices and itself contribute absorbing and possibly aggregating events coming from nearby devices. This can be used to realize composite services with a direct device-to-device coordination. This scenario is represented in the left part of Fig. 1.2.

In addition, a funnel process to be connected to an event flow spread over a multitude of devices, and/or a wide area, can be realized by initially replicating and distributing multiple actual processes, all logically part of the same distributed funnel process. Such aggregations can be used to realize services and applications in the large scale by reasoning at the level of collectives of devices and, in the end, to promote scale-independent and device-independent computations, along the lines of field-based coordination and aggregate computing approaches [9, 32, 39]. The right part of Fig. 1.2 represents such a scenario.

On the other hand, when acting at the edge level, funnel processes can be used to collect the events provided by devices at a specific location (e.g., a room, a building, or a plaza) and elaborate them to produce flows of commands to the local actuators and possibly forwarding streams of data to the cloud. Finally, nothing prevents one from associating funnel processes with some cloud servers to realize some sorts of centralized services, either explicitly within funnel process deployment specification [35] or implicitly as an opportunistic choice at the platform level. This scenario is represented in the central part of Fig. 1.2.

We emphasize that, building on previous experience on formal models of distributed and adaptive business processes [18, 19] and in distributed field-based coordination [9, 52], the Fluidware approach is meant to support a formalized description of funnel processes supporting static and behavioral property verification, independently of where the approach acts at the level of individual devices

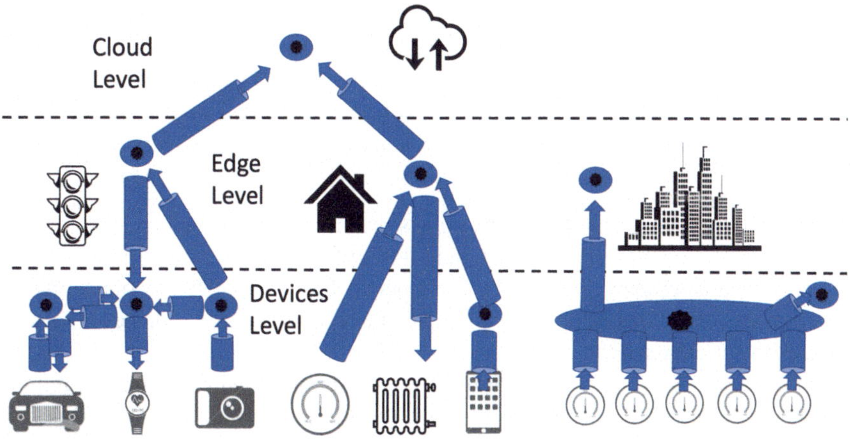

Fig. 1.2 Overview of the Fluidware approach

or at the edge/cloud level, and independently of whether it is used to implement traditional composition of services or rather in-the-large collective distributed coordination.

1.4.3 Addressing the Challenges

Let us now analyze how Fluidware has the potential to effectively address the challenges identified in Sect. 1.3.

Device Independence In Fluidware, whole collectives of IoT devices are abstracted as producers (or consumers, for actuators) of distributed streams of contextualized events, thus making system programming dependent on the availability of specific patterns of events, not on the specific identity of the devices producing or consuming them.

With reference to the case study, a filter *funnel can be put in place to get events tagged with a specific "biometric" attribute [3, 22] and can be composed (by a* pipe*) to a* relocator *to make such data be available for monitoring from a remote station owned by doctors. This will be specified without any a priori knowledge of which sensors produced such biometric data.*

Scalability In Fluidware, funnel processes can be used to access individual streams, as well as—with the same model—to aggregate, compose, and control distributed flows of events generated by myriads of devices, thus seamlessly enabling small-scale service composition and large-scale services based on collective behaviors.

With reference to the case study, a filter *funnel can be put in place to get events tagged with a specific "temperature" attribute and can be composed (by a* pipe*) to an* aggregator *to extract a stream of average values of temperature, where each event is produced each minute. This will be specified without any a priori knowledge of the dislocation and number of sensors in place: Aggregation will be transparently executed by the sensors themselves by en route combination of values [46].*

Adaptivity In Fluidware, funnel processes are not statically tied to specific IoT devices but are dynamically bound to any flows of events matching contextual and semantic characteristics. Thus, they can operate by dynamically reconnecting to different sets of devices, simply depending on their characteristics, whenever needed to react to contingencies.

With reference to the case study, a composition of funnels can be put in place to send periodic information about the house and Mark's general status to doctors. However, as soon as some physiological data exceed safety ranges for a sufficient amount of time, those funnels start matching different streams of data and automatically reconfigure to send fine-grained samples of all available physiological data to the doctors [11].

Seamless Integration of Devices, Edge, and Cloud Levels In Fluidware, funnel processes may be playing the role of redirecting streams of data to the edge, where further processes may digest them and redirect them to realize composite edge services, or they can be used to redirect streams to the cloud.

With reference to the case study, funnels like aggregators, used to perform some situation recognition across a possibly large set of devices, are specified in a way that makes it possible to carry their process on in different ways, selected depending on the available resources, or even possibly dynamically switching to address contingencies: in a fully distributed way by cooperation of devices, by having those devices simply sending their data to some cloud resource, or even by having local edge devices computing local aggregates which are sent to the cloud for further aggregation [48].

Interoperability and Security Concerning interoperability, the possibility in Fluidware to program services in a device-independent way promotes interoperability and just requires devices to host a Fluidware local proxy or—in the case of very lightweight IoT devices—to directly communicate via standard IoT protocols with Fluidware edge devices. Concerning security, Fluidware can not only rely on existing security solutions but can also integrate recent research approaches to trust computing for the IoT and wireless sensor networks [27].

Mark's house is intrinsically open and secure. First, when new kinds of sensors are deployed, they automatically become part of the Fluidware ecosystem: They either have a lightweight Fluidware software agent on board that carries on all computational activities in cooperation with others or simply send their data using standard protocols to the nearest fully featured Fluidware edge device (each room could have one installed in it). Additionally, Fluidware continuously runs processes of trust monitoring and control to detect malevolent devices and progressively make them ignored by others to prevent unwanted perturbation of the overall house—as suggested in [14].

1.5 The Path Toward Fluidware

The road-map of research activities needed to fully develop Fluidware includes:

1. Development of the Fluidware model and associated programming abstractions, suitable for defining and composing services and systems possibly involving a large number of IoT devices distributed over a large area
2. Development of a prototype platform (i.e., a middleware and associated services) to support deployment and execution of IoT systems according to the Fluidware model
3. Development of an engineering methodology to guide developers in the analysis and design of Fluidware-based complex IoT service systems and associated tools to support validation before deployment
4. Testing on IoT case studies, specifically in the area of smart health-care and AAL

1.5.1 Programming Model

From the development and programming viewpoint, it can be expected that IoT services and IoT applications will be mostly up to expert system developers. Yet, the Fluidware declarative approach will also enable local managers of a location (and, to some limited extent, end users) to be able to directly personalize the behavior of such applications with simple forms of user-level programming of funnel processes, for instance, in a similar way to "if this then that" approaches [20] or by selecting specific funnel processes and services from libraries of reusable specifications.

To formalize the Fluidware operational model and implement its basic programming interface, the following activities are to be undertaken:

- Developing the operational model of funnel processes, to serve as a blueprint for implementation of the platform, for defining composition techniques, and to check well-formedness of specifications and properties. It will need to include development of a core calculus, operational semantics, and by-construction proofs of self-stabilization and safe encapsulation.

- Implementing a library to provide the core mechanisms devised in the model, to specify and compose processes, as an interface toward the platform and existing simulators. The adoption of modern techniques can be envisaged to smoothly integrate with mainstream programming and functional-oriented declarative approaches, e.g., by the features of the Scala language as in [16].

1.5.2 Middleware

Fluidware can be supported by a middleware capable of instantiating local proxies of funnel processes and launch them in execution into the proper location [28, 30, 37] and to relocating them as needed. To promote scalability and flexibility, the platform will support interactions and coordination at three levels (as from Fig. 1.2): *direct device-to-device level* (e.g., for field-based coordination), with funnel processes directly instantiated on sets of devices, and with events flowing, aggregating and redistributing from device to device; *edge level*, with funnel processes dynamically allocated on edge computers (i.e., cloudlets or fog computers), to digest streams of events, implement local coordinated services, and possibly connect multiple edges to realize inter-edge coordination; and *cloud level*, for centralized monitoring, coordination, and storage.

The activities to develop the Fluidware middleware platform will address three related objectives:

- To define a mapping from Fluidware programming specifications into a set of distributed components to be deployed atop the Fluidware platform. Mapping will be based on a model-driven development [15] and will distinguish abstract platform-independent specifications from deployable platform-dependent components.
- To implement a distributed engine for supporting the execution of Fluidware systems and services. It will be organized as a three-layered (devices, edges, and cloud) super-peer architecture. Platform-dependent components can be dynamically activated and reconfigured in IoT devices, edge servers, and/or cloud platforms.
- To analyze interoperability and security issues. Such an activity will define guidelines for enabling devices to connect to the Fluidware platform and will analyze how a trust-oriented distributed infrastructure for inter-component security can be integrated within it.

1.5.3 Engineering Methodology

The Fluidware approach will also require the definition of new conceptual abstractions to reason about complex IoT services and applications and their requirements.

In addition, it will call for the identification of specific methodological guidelines to drive the development process, to be necessarily accompanied by Fluidware-specific tools to support the activities of the development process, and to provide correctness guarantees [31, 51].

Accordingly, there is need of synthesizing from the previously identified activities in order to make the basic engineering instruments available supporting the analysis, design, and implementation of complex IoT systems and services with Fluidware. To this end, the following activities need to be implemented.

- Identify the key conceptual abstractions which the analysis and design of Fluidware systems and services should rely on, and on this basis it will define guidelines for analysis, design, implementation, testing, and adaptation of Fluidware services and systems.
- Identify and prototype a set of tools in support of the development of Fluidware systems and services. These will include (i) basic support to verify service behavior against specifications (by model checking or static analysis) [21] and (ii) a large-scale simulator to verify overall system behaviors [15]. Prototyping of such tools will be based on extending/adapting existing techniques, tools, and simulators.

1.5.4 Application Studies

Fluidware is a general-purpose approach, suitable for developing IoT services and applications in a variety of emerging scenarios, such as smart homes, smart cities, traffic control systems, energy control systems, and smart production systems. To keep focus without losing generality, it may be proper to focus on a specific challenge related to the aging society: AAL [20]. This will enable putting the Fluidware approach at work in real-world problems, to guide its development strategies and to assess the effectiveness of the approach. In particular, we intend to test AAL applications both at the scale of a real-life domestic indoor testbed and of a large-scale simulated urban scenario. The domestic testbed will be aimed at verifying ease and expressiveness of programmability of Fluidware, as well as the effectiveness of the platform and its adaptability properties. The large-scale simulated scenario will be useful to verify Fluidware scalability and flexibility.

Given the safety-critical nature of the described scenario, many services related to, e.g., the health condition of patients or the ambient conditions of the house may be required to continuously send information to be analyzed to different actors (from doctors to caregivers). Thus, any approach for developing services and applications should more properly conceive sensors and actuators as producers and consumers of continuous flows of events, rather than as loci of services to be invoked as often happens through conventional cloud-based (virtualization) approaches.

1.6 Related Work

The Fluidware computational model definitely owns to methods proposing the simplification of distributed programming by abstracting from individual networked devices and working at the level of their collective behaviors, such as TOTA [32], SCEL [3, 22], SAPERE [52], or DECCO [12]. These also include space–time models for the universal manipulation of field data structures diffused in space and evolving with time, such as spatial [43] and aggregate computing [6, 9]. Fluidware, with its concept of funnel processes, will advance such approaches by enabling the seamless modeling of small-scale composite services, as well as large-scale collective services, and by supporting the integration of semi-decentralized approaches with fully distributed ones. Also, by exploiting the lessons of process algebras [17] and formal approaches [10], Fluidware can add a process layer which (i) internally carries on stream computation and (ii) externally defines life cycle aspects such as funnel process generation, space-time extension, interaction with environment, and deallocation. By taking inspiration from frameworks to automatically split data processing behavior for cloud- and cluster-style execution (e.g., Spark [50] and Flink [13]), Fluidware will enrich traditional collective and aggregate approaches to distributed programming by considering a transition from handling collective "fields" of data to the notion of distributed stream of events, thus addressing also nonfunctional aspects concerning the control of dynamic aspects of event generation and diffusion.

Some recent approaches to programming IoT services propose new computational abstractions as building blocks of IoT services, such as the microservices of Osmotic computing [36, 45], the core processors of EdgeIoT [44], or the deployment units of the Elastic Computing [35] and of approaches for the computing continuum [23]. These approaches share with Fluidware the idea of enabling the adaptive deployment of such building blocks at the level of both cloud and edge computers and possibly at the level of IoT devices. However, Fluidware will enrich that with an operational semantics addressing dynamically multilayered architectures, with the possibility of also acting in collective terms at the level of devices to enforce large-scale adaptive service composition, like the Opportunistic IoT Services approach based on aggregate computing [15].

Concerning middleware, a variety of platforms have been proposed to support the deployment and execution of IoT services and applications (see [37, 40] for a survey) and including solutions to adaptively handle interoperability [8, 24], context dependency [38], and adaptivity [39, 49]. Similarly to them, the Fluidware platform will promote interoperability (thanks to device independence of the funnel process abstraction), adaptivity (due to the flexible deployment of funnel processes), and context dependency (events digested by funnel processes are inherently contextual). However, it will also provide solutions for the collective execution and coordination of funnel processes on the large scale.

Concerning software engineering, it is recognized that the development of IoT systems and applications may require not only the extension of existing methods

[31] but also novel methodologies and tools [28, 51]. However, actual methodologies and tools specifically suited to future IoT scenarios are lacking, and Fluidware will fill this gap by producing novel software engineering guidelines that—although studied in the context of Fluidware—can be of general help toward the development of complex and large-scale IoT applications. Also, it will produce tools (novel or extensions of existing ones) in support of the verification of IoT system behavior and that can be of general use in the context of large-scale distributed systems.

1.7 Conclusion

In this chapter, we have presented the vision of Fluidware, an innovative approach for the development of IoT services and applications, conceived, to ease the development of flexible and robust large-scale IoT services and applications. To summarize, we can list the key innovations that Fluidware brings about.

- Its funnel process abstraction goes significantly beyond the state of the art in computational models for distributed and collective systems, allowing to declaratively express distributed processes managing contextualized streams of events in a way that is effectively scale-independent, device-independent, and independent of the temporal availability of specific devices, thus being adaptive to changes and contingencies.
- The interplay between Fluidware model and platform will fully provide utility-driven exploitation of infrastructure, enabling execution of funnel processes along the entire IoT device/edge/cloud stack, hence supporting scenarios where devices opportunistically exploit P2P interactions, edge devices, or cloud resources, to either improve performance, save energy, or speed up the sense-to-react feedback.
- The definition of novel software engineering abstractions, methodologies, and tools to support activities performed to develop Fluidware applications can shed new lights into the general software engineering issues associated with the development of complex IoT systems and services in general, independently of Fluidware.

We have developed several of these technologies in the context of the Fluidware process. These, once properly engineered, will be the basis to make Fluidware actually deliver its identified potentials.

References

1. Alrahman, Y.A., De Nicola, R., Garbi, G., Loreti, M.: A distributed coordination infrastructure for attribute-based interaction. In: Proceedings of the 38th IFIP WG 6.1 International Conference on Formal Techniques for Distributed Objects, Components, and Systems. LNCS, vol. 10854, pp. 1–20. Springer, Berlin (2018)

2. Alrahman, Y.A., De Nicola, R., Loreti, M., Tiezzi, F., Vigo, R.: A calculus for attribute-based communication. In: Proceedings of the 30th Annual ACM Symposium on Applied Computing, pp. 1840–1845. ACM, New York (2015)
3. Alrahman, Y.A., Nicola, R.D., Loreti, M.: Programming interactions in collective adaptive systems by relying on attribute-based communication. Sci. Comput. Program. **192**, 102428 (2020)
4. Andrikopoulos, V., Bucchiarone, A., Sáez, S.G., Karastoyanova, D., Mezzina, C.A.: Towards modeling and execution of collective adaptive systems. In: Proceedings of the 11th International Conference on Service-Oriented Computing. LNCS, vol. 8377, pp. 69–81. Springer, Berlin (2013)
5. Atzori, L., Iera, A., Morabito, G.: The internet of things: A survey. Comp. Netw. **54**(15), 2787–2805 (2010)
6. Audrito, G., Casadei, R., Damiani, F., Torta, G., Viroli, M.: Programming distributed collective processes for dynamic ensembles and collective tasks. In: Coordination Models and Languages—25th IFIP WG 6.1 International Conference, pp. 71–89 (2023)
7. Audrito, G., Viroli, M., Damiani, F., Pianini, D., Beal, J.: A higher-order calculus of computational fields. ACM Trans. Comput. Log. **20**(1), 5:1–5:55 (2019)
8. Ayala, I., Amor, M., Fuentes, L.: The sol agent platform: enabling group communication and interoperability of self-configuring agents in the internet of things. J. Ambient Intell. Smart Environ. **7**(2), 243–269 (2015)
9. Beal, J., Pianini, D., Viroli, M.: Aggregate programming for the internet of things. IEEE Comput. **48**(9), 22–30 (2015)
10. Belzner, L., Hölzl, M., Koch, N., Wirsing, M.: Collective autonomic systems: towards engineering principles and their foundations. In: Transactions on Foundations for Mastering Change I, LNCS, vol. 9960, pp. 180–200. Springer, Berlin (2016)
11. Bicocchi, N., Mamei, M., Zambonelli, F.: Self-organizing virtual macro sensors. ACM Trans. Auton. Adapt. Syst. (TAAS) **7**(1), 2:1–2:28 (2012)
12. Bures, T., Plasil, F., Kit, M., Tuma, P., Hoch, N.: Software abstractions for component interaction in the internet of things. IEEE Comput. **49**(12), 50–59 (2016)
13. Carbone, P., Katsifodimos, A., Ewen, S., Markl, V., Haridi, S., Tzoumas, K.: Apache Flink: stream and batch processing in a single engine. Bulletin of the IEEE Computer Society Technical Committee on Data Engineering **36**(4), 28–38 (2015)
14. Casadei, R., Alessandro, A., Viroli, M.: Towards attack-resistant aggregate computing using trust mechanisms. Sci. Comput. Program. **167**, 114–137 (2018)
15. Casadei, R., Fortino, G., Pianini, D., Russo, W., Savaglio, C., Viroli, M.: Modelling and simulation of opportunistic IoT services with aggregate computing. Future Generation Comp. Syst. **91**, 252–262 (2019)
16. Casadei, R., Viroli, M.: Towards aggregate programming in Scala. In: Proceedings of the 1st Workshop on Programming Models and Languages for Distributed Computing, pp. 5:1–5:7. ACM, New York (2016)
17. Choe, Y., Lee, M.: Algebraic method to model secure IoT. In: Domain-Specific Conceptual Modeling, pp. 335–355. Springer, Berlin (2016)
18. Cognini, R., Corradini, F., Polini, A., Re, B.: Extending feature models to express variability in business process models. In: Proceeding of Advanced Information Systems Engineering Workshops. LNBIP, vol. 215, pp. 245–256. Springer, Berlin (2015)
19. Cognini, R., Corradini, F., Polini, A., Re, B.: Business process feature model: an approach to deal with variability of business processes. In: Domain-Specific Conceptual Modeling, pp. 171–194. Springer, Berlin (2016)
20. Corno, F., De Russis, L., Roffarello, A.M.: A semantic web approach to simplifying trigger-action programming in the IoT. Computer **50**(11), 18–24 (2017)
21. Corradini, F., Fornari, F., Polini, A., Re, B., Tiezzi, F., Vandin, A.: BProVe: a formal verification framework for business process models. In: Proceedings of the 32nd International Conference on Automated Software Engineering, pp. 217–228. IEEE Computer Society, New York (2017)

22. De Nicola, R., Latella, D., Lluch-Lafuente, A., Loreti, M., Margheri, A., Massink, M., Morichetta, A., Pugliese, R., Tiezzi, F., Vandin, A.: The SCEL language: design, implementation, verification. In: Software Engineering for Collective Autonomic Systems—The ASCENS Approach, pp. 3–71. Springer, Berlin (2015)
23. Dustdar, S., Pujol, V.C., Donta, P.K.: On distributed computing continuum systems. IEEE Trans. Knowl. Data Eng. **35**(4), 4092–4105 (2022)
24. Fortino, G., Savaglio, C., Palau, C., de Puga, J., Ghanza, M., Paprzycki, M., Montesinos, M., Liotta, A., Llop, M.: Towards multi-layer interoperability of heterogeneous IoT platforms: The INTER-IoT approach. Internet of Things, pp. 199–232 (2018)
25. Fortino, G., Giannantonio, R., Gravina, R., Kuryloski, P., Jafari, R.: Enabling effective programming and flexible management of efficient body sensor network applications. IEEE Trans. Hum.-Mach. Syst. **43**(1), 115–133 (2013)
26. Fortino, G., Gravina, R.: Fall-MobileGuard: a smart real-time fall detection system. In: Proceedings of the 10th International Conference on Body Area Networks. ICST, New York (2015)
27. Fortino, G., Messina, F., Rosaci, D., Sarnè, G.M.L.: Using trust and local reputation for group formation in the cloud of things. Futur. Gener. Comput. Syst. **89**, 804–815 (2018)
28. Fortino, G., Russo, W., Savaglio, C., Shen, W., Zhou, M.: Agent-oriented cooperative smart objects: From IoT system design to implementation. IEEE Trans. Syst. Man Cybern.: Syst. **48**(11), 1939–1956 (2018)
29. Fortino, G., Re, B., Viroli, M., Zambonelli, F.: Fluidware: An approach towards adaptive and scalable programming of the IoT. In: Models, Languages, and Tools for Concurrent and Distributed Programming—Essays Dedicated to Rocco De Nicola on the Occasion of His 65th Birthday, LNCS, vol. 11665, pp. 411–427. Springer, Berlin (2019)
30. Galzarano, S., Giannantonio, R., Liotta, A., Fortino, G.: A task-oriented framework for networked wearable computing. IEEE Trans. Autom. Sci. Eng. **13**(2), 621–638 (2016)
31. Jacobson, I., Spence, I., Ng, P.W.: Is there a single method for the internet of things? Queue **15**(3), 20 (2017)
32. Mamei, M., Zambonelli, F.: Programming pervasive and mobile computing applications: The TOTA approach. ACM Trans. Softw. Eng. Methodol. **18**(4), 15:1–15:56 (2009)
33. Mariani, S., Cabri, G., Zambonelli, F.: Coordination of autonomous vehicles: taxonomy and survey. ACM Comput. Surv. **54**(1), 19:1–19:33 (2022)
34. Miorandi, D., Sicari, S., De Pellegrini, F., Chlamtac, I.: Internet of things: vision, applications and research challenges. Ad Hoc Netw. **10**(7), 1497–1516 (2012)
35. Moldovan, D., Copil, G., Dustdar, S.: Elastic systems: towards cyber-physical ecosystems of people, processes, and things. Comput. Stand. Interfaces **57**, 76–82 (2018)
36. Nardelli, M., Nastic, S., Dustdar, S., Villari, M., Ranjan, R.: Osmotic flow: osmotic computing+ IoT workflow. IEEE Cloud Comput. **4**(2), 68–75 (2017)
37. Palade, A., Cabrera, C., White, G., Razzaque, M.A., Clarke, S.: Middleware for Internet of Things: A Quantitative Evaluation in Small Scale, pp. 1–6. IEEE, New York (2017)
38. Perera, C., Zaslavsky, A., Christen, P., Georgakopoulos, D.: Context aware computing for the internet of things: A survey. IEEE Commun. Surv. Tutorials **16**(1), 414–454 (2014)
39. Pianini, D., Casadei, R., Viroli, M., Mariani, S., Zambonelli, F.: Time-fluid field-based coordination through programmable distributed schedulers. Log. Methods Comput. Sci. **17**(4) (2021)
40. Picone, M., Mamei, M., Zambonelli, F.: A flexible and modular architecture for edge digital twin: Implementation and evaluation. ACM Trans. Internet Things **4**(1), 8:1–8:32 (2023)
41. Pournaras, E.: Overlay service computing—modular and reconfigurable collective adaptive systems. Scalable Computing: Practice and Experience **16**(3), 249–270 (2015)
42. Rausch, T., Dustdar, S., Ranjan, R.: Osmotic message-oriented middleware for the internet of things. IEEE Cloud Comput. **5**(2), 17–25 (2018)
43. Shekhar, S., Feiner, S.K., Aref, W.G.: Spatial computing. Commun. ACM **59**(1), 72–81 (2016)
44. Sun, X., Ansari, N.: EdgeIoT: Mobile edge computing for the internet of things. IEEE Commun. Mag. **54**(12), 22–29 (2016)

45. Villari, M., Fazio, M., Dustdar, S., Rana, O., Ranjan, R.: Osmotic computing: A new paradigm for edge/cloud integration. IEEE Cloud Comput. **3**(6), 76–83 (2016)
46. Viroli, M., Audrito, G., Beal, J., Damiani, F., Pianini, D.: Engineering resilient collective adaptive systems by self-stabilisation. ACM Trans. Model. Comput. Simul. **28**(2), 16:1–16:28 (2018)
47. Viroli, M., Casadei, M., Montagna, S., Zambonelli, F.: Spatial coordination of pervasive services through chemical-inspired tuple spaces. ACM Trans. Auton. Adapt. Syst. **6**(2), 14:1–14:24 (2011)
48. Viroli, M., Casadei, R., Pianini, D.: On execution platforms for large-scale aggregate computing. In: Proceedings of the International Joint Conference on Pervasive and Ubiquitous Computing, pp. 1321–1326. ACM, New York (2016)
49. Vlacheas, P., Giaffreda, R., Stavroulaki, V., Kelaidonis, D., Foteinos, V., Poulios, G., Demestichas, P., Somov, A., Biswas, A.R., Moessner, K.: Enabling smart cities through a cognitive management framework for the internet of things. IEEE Commun. Mag. **51**(6), 102–111 (2013)
50. Zaharia, M., Xin, R.S., Wendell, P., Das, T., Armbrust, M., Dave, A., Meng, X., Rosen, J., Venkataraman, S., Franklin, M.J., et al.: Apache spark: a unified engine for big data processing. Commun. ACM **59**(11), 56–65 (2016)
51. Zambonelli, F.: Key abstractions for IoT-oriented software engineering. IEEE Softw. **34**(1), 38–45 (2017)
52. Zambonelli, F., Omicini, A., Anzengruber, B., Castelli, G., Angelis, F.L.D., Serugendo, G.D.M., Dobson, S.A., Fernandez-Marquez, J.L., Ferscha, A., Mamei, M., Mariani, S., Molesini, A., Montagna, S., Nieminen, J., Pianini, D., Risoldi, M., Rosi, A., Stevenson, G., Viroli, M., Ye, J.: Developing pervasive multi-agent systems with nature-inspired coordination. Pervasive Mob. Comput. **17**, 236–252 (2015)

Chapter 2
Programming Approaches for Large-Scale IoT System Development: State of the Art

Roberto Casadei, Fabrizio Fornari, Stefano Mariani, and Claudio Savaglio

2.1 Introduction

Various methods exist for programming IoT systems, often involving ad hoc combinations of software libraries, frameworks, and platforms. This can lead to confusion about the best approaches and how to integrate them effectively [25, 99].

The intrinsic complexity of IoT programming at scale is due to the heterogeneity of devices, their distribution, and the local-to-global connection [23, 66]. To address this complexity, IoT programming research focuses on three fundamental facets: how to manage the perception of properties of interest from the IoT environment through sensors (*sensing* facet), how to process this information for decision-making to achieve system goals (*processing* facet), and finally how to enact these decisions by affecting the environment through actuators (*actuation* facet). Accordingly, IoT programming research spans the entire software stack: languages and libraries for programming devices, platforms supporting data streaming and event processing, frameworks for distributed computing, and service deployment.

R. Casadei (✉)
University of Bologna, Cesena, Italy
e-mail: roby.casadei@unibo.it

F. Fornari
University of Camerino, Camerino, Italy
e-mail: fabrizio.fornari@unicam.it

S. Mariani
University of Modena and Reggio Emilia, Reggio Emilia, Italy
e-mail: stefano.mariani@unimore.it

C. Savaglio
Università della Calabria, Rende, Italy
e-mail: csavaglio@dimes.unical.it

F. Zambonelli et al. (eds.), *Fluidware*, Internet of Things,
https://doi.org/10.1007/978-3-031-62146-8_2

Several surveys have identified the most active research trends, categorizing the driving application domains and emphasizing open issues and opportunities for further research [69, 75, 81, 87, 90]. However, these surveys tend to focus on a particular approach (e.g., stream processing but not macro-programming), infrastructural layer (e.g., the cloud but not the Edge and platforms but not languages), and/or feature set (e.g., opportunistic deployment but not adaptiveness), failing to clarify how different approaches fit into the overall picture. Therefore, an integrative analysis of the IoT programming research landscape is needed.

For these reasons, we provide a meta-review (i.e., first targets of our review are surveys themselves) of the literature about "IoT programming." Our goal is to address the following questions: (i) what are the research areas contributing the most to IoT programming? (ii) how do they cover the sensing, processing, and actuation facets? (iii) what are the challenges yet to be dealt with?

2.2 Conceptual Framework and Review Method

Every IoT system has three distinct but closely related facets of computation:

- The *sensing facet* is concerned with how to *perceive* the "state of the world" according to the application goals. This facet includes gathering measured data from sensor devices, preprocessing such data to derive more complex situations [77], and possibly even putting such situations into causal relations [78].
- The *processing facet* is concerned with elaborating available data to *decide* which actions to carry out to achieve the system goals. This facet includes simple reactive rule engines producing commands depending on the current situation [48], more complex forms of event processing, agent-oriented computing to have goal-oriented behaviours [86], and so on.
- The *actuation facet* is concerned with how to *affect* the state of the world according to the application goals. This facet includes low-level programming of individual devices with mainstream languages, collective engineering of the global system behaviour, and many other different approaches [16].

We place the surveyed literature along these facets to clarify which are most considered and which require further development.

To conduct our survey, we began by performing keyword-based searches on Scopus and DBLP, focusing on papers that were already surveys themselves. We performed several searches including "IoT/CPS" and "programming," such as "IoT programming languages survey," "CPS programming approaches review," and similar combinations. The reason for this methodology is that the topic of IoT programming is so broad and has so many different nuances that it would be impossible to rigorously scrutinize all the available literature. By focusing on existing surveys, we can realistically analyse them in detail and later perform snowballing (i.e., searching for therein referenced papers) through the most relevant contributions found.

The results obtained are further filtered:

1. By comparing title and metadata, we filtered out duplicates and different versions of same paper (e.g., conference and journal version).
2. By reading the abstract, we filtered out papers that clearly were out of scope (e.g., no focus on programming).
3. By reading the full paper, we filtered out papers that were superseded by another one, sensibly overlapped with another one, out of scope, etc.

We then grouped the collected papers into categories to snowball and get a more focused overview of the research landscape.

Stream processing (SP) Babcock et al. [12] is a computing paradigm that focusses how to gather, process, and analyse continuous flows of data called *streams* within a small time period from the time of receiving. SP gained popularity within the IoT landscape as often times IoT deployments need first and foremost to tackle the challenge of collecting and processing timely vast amounts of data continuously coming from some system under monitoring, as in the case of Industry 4.0, smart supply chain and logistics, intelligent transportation, and telemedicine. As a consequence, SP became one of the fundamental assets available to developers for programming IoT systems, especially covering the "sensing facet" of our conceptual framework.

Complex Event Processing (CEP) Luckham [59] builds on the abstraction of low-level event to create a new high-level one based on temporal, spatial, or causal relations. CEP is closely related to SP: It shares the general idea of processing potentially *unbounded* streams of data and many techniques for actually doing so. CEP comes into play for IoT applications not limited to simple monitoring tasks—for which SP is often enough. Whenever there is the need for complex analytics and insights, and to some extent of closing the feedback loop by controlling the monitored system back, CEP has an edge over other approaches, thanks to its native capability of detecting complex patterns over incoming events data and of triggering business processes reactively. For this reason, we put CEP at the "processing facet" of IoT programming.

Micro- and Macro-Programming (MMP) . This category includes approaches not belonging to a well-defined research area, but spanning from traditional programming (micro) where a programme is developed with a specific, *single* device in mind (e.g., a computer, a mobile phone, and an embedded device), to emerging programming paradigms that target a *multitude* of devices at once (macro) [42], by developing a global, high-level programme that is then "translated" into lower level, device-bound (traditional) programmes [16]. A description of each approach is given in the corresponding Sect. 2.5. Being these approaches focussed on *controlling* devices programmatically, they mostly belong to the "actuation facet" of our conceptual framework.

We then expanded our original keyword-based search with terms directly stemming from the identified research areas, in a snowballing process. On the results retrieved, the same filtering steps already mentioned have been applied. The papers survived

Table 2.1 Main SP and ESP surveys, most recent first

Ref.	Brief description	Year	Category
[97]	Focus on techniques for Cloud elasticity (mostly QoS-aware resource usage) at operator placement/execution level	2021	DOM
[37]	Historical perspective, emphasizing evolution of research landscape, and comparing different SP platforms generations	2020	GEN
[83]	Systematic analysis of strengths and weaknesses of open source SP techs for predictive maintenance in Industry 4.0	2020	DOM
[46]	Comparative study of selected distributed SP and analytics platforms	2019	GEN
[67]	Evaluation of selected open source distributed SP platforms deployed at the "data layer" of Smart City applications	2019	DOM
[95]	Review of scheduling techniques for distributed SP	2019	SPEC
[82]	Focus on techniques for parallelization and elasticity of SP	2019	SPEC
[33]	Focus on SP mechanisms leveraging resource elasticity in the Cloud	2018	SPEC
[30]	Overview of ESP architectures, use cases, and open research issues	2018	GEN
[104]	Specifically concerned with real-time SP in IoT deployments, hence surveys-related techniques, challenges, and technologies	2016	SPEC

are actually the one cited throughout this whole paper and summarized in Tables 2.1, 2.2, 2.3, 2.4, and 2.5.

It is worth emphasizing a few aspects about these research areas: (i) First, all of them are increasingly used in the context of IoT services and applications, either as components of a broader architecture or in isolation, as key enablers of programming tasks related to monitoring, decision-making, and distributed control, and (ii) second, SP mostly deals with the sensing facet, and partially with the processing one, CEP mostly deals with the processing facet, and partially with the sensing and actuation ones, and MMP is primarily concerned with the actuation facet and partially with the processing one; hence they seemingly fit our conceptual framework and altogether cover it broadly (see Sects. 2.3, 2.4, and 2.5 for a discussion of each area). On the one hand, this means that each research area brings effective solutions to specific issues of programming large-scale, autonomous IoT systems, and on the other hand, that these solutions are only partially solving the problem; hence integrating them is likely the key for better IoT programming.

The surveys and papers collected have been roughly divided into three categories—Tables 2.1, 2.3, and 2.5, one per each macro-area—depending on their scope and intended goal: A first category (GEN) is focussed on comparison of available languages and platforms across many **gen**eral criteria, a second one (DOM) is concerned with the usage of languages and platforms in specific application **dom**ains (e.g., Smart Cities and Industry 4.0.), and a third one (SPEC) examines a **spec**ific design choice or property (e.g., main abstraction, scheduling techniques, and parallelization) of the languages and platforms and compares them based on the degree of support provided.

Table 2.2 Summary of the SP main characteristics

	Flink [19]	Heron [55]	Kafka streams [52]	Samza [71]	Spark streaming [106]	Storm [98]
Programming model	Declarative, Imperative	Imperative	Declarative	Compositional	Declarative, Imperative	Imperative, Compositional
Processing model	Hybrid	Streaming	Batch/Streaming	Hybrid	Micro-Batch	Streaming, Operator Based, Hybrid (with Trident)
Deployment	Cluster, Cloud, Fog, Local	Cluster, Cloud, Fog	Containers, VMs, Cloud, Edge	Cluster Standalone Public-cloud Containers Bare-metal	Cluster, Cloud, Fog	Cluster, Cloud, Fog
Parallelization	Task, Data	Task, Data	Task, Data	Task, Data	Task, Data	Task, Data
Stream Primitive	DataStream	Tuple	Stream	Message	DataSet DataFrame	Tuple
Delivery guarantee	Exactly once	At least once, Effectively once	At least once, At most once, Exactly once	At least once	Exactly once, At-least-once	At least once, Exactly once (with Trident)
Scalability	Horizontal Vertical	Horizontal Vertical	Horizontal Vertical	Horizontal Vertical	Horizontal Vertical	Horizontal Vertical
Data flow abstraction	DAG	DAG	DAG	DAG	DAG	DAG
Fault tolerance	Stream replay Checkpoint	Checkpoint	Stream replay Checkpoint	Incremental Checkpointing	Checkpoint	Acking, Checkpoint, Stream replay, Highly fault-tolerant
API languages	Java, Scala, Python, SQL	C++, Python Java, No SQL	Java, Scala, KSQL	Java, SQL, Python	Java, Scala, R, Python, Spark SQL	Java, Python, SQL

Table 2.3 Main CEP surveys, most recent first

Ref.	Brief description	Year	Category
[38]	Focus on CER (subtask of CEP), especially regarding scale-out mechanisms for query execution	2020	SPEC
[108]	Focus on out-Cloud deployment, in particular adaptations needed for CEP platforms to meet the Fog layer	2020	SPEC
[107]	Overview on the state of the art of CEP, also discussing the relationship with SP to improve efficacy in IoT deployments	2020	GEN
[6]	Focus on probabilistic CEP for uncertainty handling	2017	SPEC
[64]	Specifically concerned with application of CEP to healthcare, with a particular focus on how to exploit healthcare sensor devices	2017	DOM
[8]	Focus on languages for CEP applied to multimedia sensor networks	2016	SPEC

2.3 Stream Processing

SP is a programming paradigm that sees application development as processing *continuously* incoming data: Given a sequence of data (a stream), a series of operations is applied to each element in order. When data is associated with a timestamp, the literature usually changes terminology and talks about Event Stream Processing (ESP) platforms. (E)SP deals with high volumes of data produced and organized in a single *stream*, ordered by timestamps, that have to be processed as fast as possible to provide near real-time insight. The typical example is a stock market feed that should be analysed to determine whether to buy or sell a stock in such a way to gain profit. ESP applications do not normally include features leveraging event causality or event hierarchies but can do so [61].

Amongst the characteristics that an SP platform should present [93], a dominant one is the ability to process data on-the-fly, which can be regarded as a form of *situatedness* along the temporal dimension. Especially in near-real time systems, waiting for data to be stored (e.g., in a database) before being processed adds unnecessary latency. Modern SP platforms are also capable of partitioning computational load and *scaling* applications automatically and of distributing their processing needs across multiple processors and machines. In the most advanced systems, this distribution is handled automatically and transparently, as a form of *opportunistic deployment*.

SP dates back to projects such as Aurora/Borealis [2, 3], Stream [9], and TelegraphCQ [22]. Nowadays, the most popular SP platforms are Flink, Heron, Kafka Streams, Samza, Spark Streaming, and Storm (all referenced in the following). Several contributions have reported a comparison of these platforms in terms of the functionalities and applications domain they support: We provide a summary of such literature to quickly "frame" the research area and to later emphasize which characteristics of SP platforms do matter the most for our conceptual framework.

Table 2.4 Summary of the main characteristics of the surveyed CEP systems (rows and columns are exchanged w.r.t. Table 2.2 to improve layout)

	Filtering	Prioritization	Patterns	Exceptions	Process triggering	Event hierarchies	Deployment
SASE [103]	Event Algebra		Active Instance Stack (AIS) for events sequence construction on NFA			Primitive types to build complex types	Centralized
Esper [43]	Event Processing Language	Priority language construct to control processing order	Pattern matching via regular expressions + event correlation + control on patterns execution (e.g., repeat-until, every-distinct, . . .)			Event-type inheritance and polymorphism for hierarchies	Processing is centralized data acquisition distributed
Siddhi [94]	Siddhi Streaming SQL	Via rabbitmq's AMQP message priority	Pattern and sequence queries via state machines				Centralized as library + microservice on Docker Kubernetes
Cayuga [31]	Automata-based	Priority queues	Sequences of correlated events + "safety conditions" between consecutive events + patterns composition			Via query chains	Centralized

(continued)

Table 2.4 (continued)

	Filtering	Prioritization	Patterns	Exceptions	Process triggering	Event hierarchies	Deployment
CEP [5]	DOLCE language		Causal and temporal patterns using predefined rules based on thresholds + dynamic rules				Distributed on embedded devices
Oracle CEP [73]	Oracle CQL		`CQL MATCH_RECOGNIZE` condition		POJOs triggered by events	Event Processing Networks to create hierarchy of processing agents and events	Clustered
OpenCEP [44]	Automata and tree-based	Consumption policies and selection strategies	Patterns require structure (sequence, and, etc.) + conditions on each item + time window				Standalone library + Cloud deployment + integration with SP platforms
NextCEP [89]	Automata-based		SQL-like language + filter, union, iteration, time, …)	Dedicated language operator			Clustered
TIBCO Business Events [40]	Production Rules		Sequencing, duplicate detection, and others		Event-driven process orchestration		Distributed via Docker or Kubernetes + TIBCO Cloud + on premise
EdgeCEP [24]	Specification language for relational expressions		By stating an explicit event key and contents of interest				Edge

BoboCEP [76]	Automata-based		Based on NFA				Distributed on Edge
Decision CEP Engine [45]	Data stream query language						Clustered
StreamInsight [7]	Declarative LINQ						Centralized
Drools Fusion [96]	Production Rules	Rules can have priorities	Improved Rete algorithm + temporal operators		BRMS to trigger business process execution		Centralized
ILOG JRules [80]	Production Rules				BRMS to trigger and support business process execution		Centralized
Spark CEP [85]	Data stream query language						Distributed
Proton [91]	Data stream query language		Filtering + join for multiple events + sequencing + aggregation functions + track values in time			Event Processing Networks to create event hierarchies	Centralized + distributed via STORM
EasyFlinkCEP [39]	Extended regular expressions		EasyFlinkCEP Operator + FlinkCEP's language to define CEP queries + time windows				Clustered

NFA = Nondeterministic Finite Automata, BRMS = Business rules management system

Table 2.5 Main MMP surveys, most recent first

Ref.	Brief description	Year	Category
[20]	Survey about macro-programming approaches, whatever the application domain	2023	SPEC
[32]	Broad view on the development of IoT systems from the standpoint of software engineering practice	2022	GEN
[47]	Macro-programming, but specifically for the IoT domain	2021	DOM
[54]	Visual programming approaches not only in IoT, but education and robotics too	2021	SPEC
[13]	IoT development platforms, with emphasis on communication and security	2021	GEN
[36]	Review of IoT platforms from an architectural perspective only	2021	SPEC
[84]	DSLs for used in IoT focus on evaluation methods	2021	SPEC
[58]	Enabling technologies for asynchronous and real-time programming of safety-critical IoT and CPS	2019	SPEC
[17]	Asynchronous programming for IoT and embedded systems	2019	GEN
[79]	High-level study on 13 visual IoT programming languages	2017	GEN
[53]	Focus on distributed programming languages and tools for embedded systems within the IoT ecosystem	2016	SPEC
[74]	Comprehensive features' analysis and classification (node-centric, database, macro, model-driven) of IoT programming approaches	2015	GEN
[92]	Low-level languages (both mainstreams and dialects) for CPS and embedded systems	2015	GEN
[101]	Overview of conventional composition mechanisms for CPSs (monolithic, object-oriented, modular, component-based, service-oriented	2010	SPEC

2.3.1 Related Surveys

Many surveys about SP or ESP can be found, witnessing the huge interest in the topic, as summarized in Table 2.1.

2.3.2 Main Characteristics

Here, we derive from the surveyed literature the main characteristics of SP platforms (Table 2.2) and compare well-known SP platforms based on the degree of support they provide to those characteristics.

- *Programming Model.* It defines if programmers use *declarative* queries, *imperative* programming, or *compositional* programming to specify the operations or execution patterns of an SP system. Imperative programming specifies how to generate data structures, how to store temporary intermediate results within these structures, and how to convert these intermediate results into final results.

Declarative programming defines what tasks to complete instead of delivering as many details as imperative programming does. Compositional approaches offer basic building blocks for composing custom operators and topologies.

- *Processing Model.* Whether data undergo *batch* processing, *micro-batch* processing, or *native* stream processing (tuple by tuple). Batch processing stores data first and then applies processing; micro-batch processing runs batch processing on few data clustered together (the micro-batch) so that processing occurs more frequently; native stream processing collects and processes data immediately, as they are generated.
- *Deployment.* It is the infrastructure on which an SP system is deployed, e.g., a cluster, the Cloud, Fog, Edge, etc.
- *Parallelization.* Parallel SP reduces queuing latency and increases throughput, as it allows processing multiple streams simultaneously instead of sequentially. With *task* parallelization, multiple operations (i.e., tasks) can run in parallel on the same stream. With *data* parallelization, multiple instances of the same operator can act in parallel on different streams.
- *Stream Primitive.* It refers to the first-class data structure of the streaming system. In Kafka Streams,[1] for instance, it is an ordered, replayable, and fault-tolerant sequence of immutable data records, where a data record is defined as a key–value pair. In Samza,[2] instead, the stream primitive is a message, processed one at a time upon receiving. Finally, in Spark Streaming[3] Datasets and DataFrames are used, where a Dataset is a distributed collection of data, and a DataFrame is a Dataset organized into named columns—hence conceptually equivalent to a table in a relational database.
- *Delivery Guarantee.* The kind of warranty that data will be delivered from the SP platform to its client applications: at most once, at least once, and effectively once (or exactly once). With *at most once* data will be delivered once or not at all. With *at least once* data is guaranteed to be delivered, but possibly more than once. With *effectively once* data will be delivered exactly once.
- *Scalability.* The kind of scalability operations supported, e.g., adding computational resources to a processing unit (vertical) vs. adding processing units (horizontal).
- *Data Flow Abstraction.* It is the abstraction used to handle flow of data between stream operators. Most SP platforms adopt a *Directed Acyclic Graph* (DAG) of operators to structure the processing application in topologies.
- *Fault Tolerance.* The ability of an SP system to tolerate failures. Data loss and resource access loss are common failures of distributed SP engines, and common recovery strategies are based on replaying the whole data stream or restore the engine state to a checkpoint (before the fault) to repeat the processing pipeline only from there.

---

[1] https://kafka.apache.org/11/documentation/streams/core-concepts

[2] http://samza.apache.org/learn/documentation/1.6.0/core-concepts/core-concepts.html

[3] https://spark.apache.org/docs/latest/sql-programming-guide.html

- *API Languages*. The programming paradigm and actual language stack supported. The trend is to integrate with mainstream, general purpose languages, rather than to provide custom domain-specific languages.

In Table 2.2 the design choices of each specific SP platform (columns) to implement the corresponding characteristic (rows) are briefly described by keywords.

2.3.3 *Summary*

As regards the computation facets (vertical slices), SP is mostly concerned with sensing as it is usually adopted to gather sensor measurements in IoT deployments. However, some SP platforms and frameworks also offer processing techniques and algorithms; hence the processing facet is also partially covered.

2.4 Complex Event Processing

Complex Event Processing (CEP) systems are concerned with processing streams of *events*, that is, data describing situations and happenings in a system under observation [59]. Specifically, they target the definition and detection of high-level situations of interest, or *composite events*, starting from streams of *primitive events* [29]. Event Stream Processing (ESP), as an evolution of SP, and CEP actually indicate different ways to manage events [60]: ESP is a subset of CEP, with very fast implementations available, that can handle many thousands of events per second, at the price of lower processing capabilities. Nowadays, an increasing number of enterprises are slowly progressing from ESP towards CEP [28, 50, 62]. CEP foremost goal is to enable abstraction of low-level events into a new high-level event based on temporal, spatial, or causal patterns. The patterns to be detected are generally specified by domain experts, and the new events generated from pattern matching are viewed as being at a higher level of abstraction than those in input. As such, CEP allows for a stepwise abstraction process, easing the increasing flood of events that modern information systems have to deal with.

CEP is usually traced back to projects such as Amit [4], PADRES [57], SASE [41, 103], PEEX [49], RACED [26], and TESLA/T-Rex [27].

2.4.1 *Related Surveys*

As for SP, most of the collected surveys can be divided into three categories, as depicted in Table 2.3: some focus on a generic comparison of the Complex Event Processing systems (GEN), others focus on the usage of CEP in specific application domains (DOMs), and others focus on specific characteristics or mechanisms of the CEP systems, such as querying techniques, parallelization, and uncertainty handling (SPEC).

2.4.2 Main Characteristics

Table 2.4 reports a comparison similar to the one reported in Table 2.2, but in the context of CEP—please notice that columns and rows are inverted w.r.t. Table 2.2 to preserve readability.

- *Filtering*. The ability to discard irrelevant events, so as to reduce the amount of events to be considered for processing, is of paramount importance to keep CEP systems performant and focussed on the application goals. The criteria used for filtering are usually defined at design-time, but it is desirable to have ways to at least adjust them at run-time.
- *Prioritization*. CEP platforms must be able to establish which events to process sooner than others, since they may require immediate actions that critically affect the business. Again, the criteria for establishing the priority order are usually statically defined at design-time but are desirable to have ways to adapt such priorities during run-time, as a reaction to the system dynamics.
- *Patterns*. Detection and matching of event patterns is what essentially defines CEP itself. Such patterns are usually defined at design-time and cannot be altered, as the automatic mechanisms of events handling underlying CEP would be compromised. Pattern detection, matching, and querying are complex tasks that are usually exposed to the system designer through a suitable API constituting the *Event Query Language (EQL)*, further discussed below. This is one crucial difference between CEP and SP.
- *Exceptions*. Within CEP, exception handling refers to a pattern that fails to find any match within a specific time period. This may indicate the presence of some kind of anomaly, and hence an exceptional event should be generated, to undergo CEP on its own.
- *Process triggering*. Patterns of events can be used to trigger business processes that react to expected patterns with the automatic execution of predefined business actions. This is another notable difference between CEP and SP, as in the latter case process triggering is entirely responsibility of the application developer.
- *Event hierarchies*. The capability of building layers of events at different levels of abstraction, automatically shaped by the CEP platform, and linked together so as to preserve/restore/build causality.
- *Deployment*. How and where CEP solutions are deployed: Centralized, Distributed, Clustered, Cloud, Edge, etc.

As pattern detection and matching is a cornerstone of CEP, it is necessary to expand on it. What enables system designers to define pattern matching is the EQL, a high-level programming language for querying events [35, 38]. Several types of EQL have been defined in the CEP literature:

- In *event algebras*, complex event queries are expressed by composing single events using different composition operators, such as conjunction of events (all events must happen, possibly at different times), sequence (all events happen in the specified order), and negation within a sequence (an event does not happen in the time between two other events).
- *Data stream query languages* have been developed in the context of relational data stream management systems; hence they are usually based on SQL. They are applied in situations where loading data into a traditional database would require too much time, and therefore they target near real-time applications.
- *Production rules* are not an EQL strictly speaking; however they offer a convenient and flexible way of implementing EQL in the form of IF <Condition> THEN <Action> rules and hence are largely used in business rules management systems.
- *Logic languages* express event queries in logic-style formulas [10]. An early representative of this language style is the event calculus [51], which has been used to model event querying and reasoning tasks in languages such as Prolog. Logic languages have strong formal foundations and allow an intuitive specification of complex temporal conditions.
- Finally, in *tree-based* approaches patterns are represented as a tree structure, where leaves are primitive events and internal nodes are the operators contributing to define the pattern.

2.4.3 Summary

CEP somewhat encompasses all the three facets of computation, although rather weakly along the actuation one. With respect to SP, thus, we are moving from (almost) exclusively dealing with the sensing facet of IoT programming, to including more substantially the processing facet into the concerns of IoT developers. This is mostly due to the general capability of CEP to support more comprehensive and advanced forms of processing on input data (events) w.r.t. SP: Advanced pattern matching techniques and event query languages, together with the capability to express business process triggering rules and policies, make CEP much more suited to program processing pipelines, focussing not only on aggregation of sensory data streams.

2.5 Micro- and Macro-programming

In the IoT, a common goal is to programme the *global* or *collective* behaviour of the entire system or groups of devices [21]. This often entails coordination and information exchange amongst the devices, so that the overall action is coherent and informed (i.e., based on purely local information). In the literature, it is possible to distinguish between two classes of approaches [20]: informally, *micro* approaches that "take the perspective" of individual devices and address the behaviour of specific roles within the system, and *macro* approaches that somewhat abstract multiple devices into a "single conceptual machine" that is the actual target of programming (*macro-programming* [20]). To the best of our knowledge, the term macro-programming has started spreading at the beginning of 2000s in the context of research on programming of wireless sensor networks [68]. Micro-programming includes traditional approaches, e.g., based on mainstream programming languages, where the programmer specifies a script of instructions to control passive elements or defines the control loop of each individual active device. Macro-programming, instead, includes approaches that, e.g., abstract a whole network of devices as a database (cf. TinyDB [63]) or as a *spatial computing* system [15]; a notable macro-programming approach is given by the so-called *aggregate computing* paradigm [100].

2.5.1 Related Surveys

To the best of our knowledge, the most comprehensive surveys that *explicitly address* IoT programming languages and frameworks are provided in the book chapter by Krishnamurthy et al. [53] and in a specific section of the survey by Dias et al. [32]—see Table 2.5. The former reviews programming languages specifically for embedded systems (including various flavours of C), message passing technologies (RPC flavours, COAP, and MQTT), coordination languages (Linda, eLinda, Orc, and Jolie) and discusses the notion of polyglot programming. The latter, instead, provides a much broader overview over the whole software engineering practice of developing IoT systems. In particular, the overview about development tools provides many useful insights about programming languages and frameworks, such as the kind of programming paradigm promoted (e.g., visual vs. model-driven), that are partially overlapped with our contribution (e.g., they do analyse support to distributed/decentralized programming). However, such surveys do not help researchers and practitioners in understanding which one of the many different programming tools is most suited for different "tasks" in IoT system development.

A more similar work to ours, if restricted to the micro- vs. macro-categorization, is by Patel et al. [74], which classifies IoT programming models into four categories: (i) *Node-centric programming*, where the nodes of an IoT system are managed individually (i.e., micro-programming) (ii) *Database approach*, where the IoT system is abstracted as a database (iii) *Macro-programming*, where abstractions targeting multiple nodes at once are considered and (iv) *Model-driven development*, where multiple programming perspectives are addressed at once Also, two surveys specifically targeted at macro-programming have recently been published [20, 47]. They both do an excellent job in sorting out the different existing approaches into a taxonomy.

2.5.2 Main Characteristics of MMP

Here follows an overview of the main aspects along which Micro-/Macro-Programming (MMP) approaches can be categorized, also reported in Table 2.6. In line with previous sections, this overview is needed to analyse whether MMP approaches are mostly dealing with the sensing, processing, or actuation facet of IoT programming.

- *Scope Micro-programming* (or, node-centric) considers one individual device at a time. Such a bottom-up approach, adopted by, e.g., FRASAD [70], COMPOSE [34], NODE-RED [1], and mainstream programming languages such as Java, C, and Python, requires to specify, for each node, its input data, its processing functions, and its actuation commands. Conversely, recent *macro*-programming approaches such as DDFlow [72], PyoT [11], D'Artagnan [65], SmartSociety platform [88], and EdgeProg [56], aim at enabling programming of an IoT system as a whole, with a global (or *collective*) perspective.
- *Main Abstraction*. For cloud-based IoT systems, the most recurrent abstraction is the "Service" one: For instance, in COMPOSE [34] behaviours can be modelled as RESTful services. Node-centric IoT systems, instead, mostly leverage on *flow-based* programming abstractions, as networks of "black box" processes communicating through data chunks travelling across predefined connections (or, *wires*). Many are instead the abstractions used for macro-programming, mostly due to the field being relatively new, hence still fragmented amongst the different approaches put forward by different research groups. Some examples are *ensembles* (e.g., [100]), i.e., dynamic groups of devices sharing a goal, *collective interfaces* (e.g., [72]), namely mechanisms to address or refer to a dynamic number of recipients or senders, and *collective tasks* (e.g., [88]) that are operations meant to be carried out collaboratively by a group of devices.
- *Programming Model*. Imperative programming is the most used one, by both micro- and macro-level programming frameworks. Rule-based programming models, such as FRASAD [70], allow defining local behaviours of sensor nodes through a set of flexible and extensible rules. A specific nuance of such models is the visual programming model that allows easy implementation of Event-Condition-Action (ECA) rules [14]. COMPOSE is quite peculiar as it allows expressing the behaviour of IoT devices by means of basic logical, string, and arithmetic operators specified through a JSON document. Programming models for collective behaviours typically rely on a single declarative specification dictating how several distributed components should act and interact. For example, aggregate programming [16] is a functional, declarative approach for programming the self-organization logic of networked IoT systems from a global perspective.
- *Architectural/deployment view*. The majority of the surveyed works follow a standard two-tier architecture where local devices "sense and forward" to the Cloud. Only NodeRed, JAMScript, and EdgeProg, with the introduction of an intermediate Edge layer, allow also local data processing, shifting to a three-tier

Table 2.6 Summary of the MMP main characteristics

	Scope	Main abstraction	Programming model	Architectural/deployment view	Execution Model
FRASAD [70]	Individual	Reaction Rules, Messages	Visual block language	Local WSN	Reactive
SMART-BLOCK [14]	Individual	Reaction Rules	Visual block language	Cloud-based	Reactive
COMPOSE [34]	Individual	Smart Objects, Resource (REST)	Imperative, object-oriented DSL	Cloud-based	Customizable
Node-Red [1]	Individual	Data Flow	Visual block language	Edge-to-Cloud	Customizable
JAMScript [102]	Individual	Nodes, Activities	Imperative, object-oriented DSL	Edge-to-Cloud	Reactive
PyoT [11]	Collective	Resource (REST)	Imperative, object-oriented, macro-programming DSL	Master-worker	Reactive
DDFlow [72]	Collective	Data Flow	Visual block language, macro-programming	Master-worker	Reactive
D'Artagnan [65]	Collective	Data Stream, Data Flow	Functional, stream-based, macro-programming DSL	Logical network any deployment	Reactive
SmartSociety [88]	Collective	Collective task	Object-oriented DSL	Cloud-based	Customizable
EdgeProg [56]	Collective	Reaction rules, Virtual sensors	Rule-based DSL	Edge server + IoT nodes	Reactive
Aggregate Computing [100]	Collective	Computational Field	Functional declarative macro-programming	Logical network any deployment	Self-organizing

architecture. D'Artagnan and Aggregate Computing, instead, promote a different approach: They enable to programme logical networks of devices and operations, abstracting from the underlying middleware and physical systems and hence allow expressing IoT systems in a deployment-transparent manner. In such a way, they pave the way to a full and opportunistic exploitation of IoT resources, from the edge up to the cloud, with obvious benefits in terms of QoS and QoE.
- *Execution Model.* Most commonly system execution is either *time-driven*, *event-driven*, or *reactive*, which is reasonable, since activity may be triggered by new sensor data or local events. In WSN systems, event-driven and time-driven routing protocols are the main options [105]. Macro-programming approaches typically build on reactive, rather than fully proactive, execution models [20, 47], the latter being covered mostly in agent-oriented paradigms [18].

2.5.3 Summary

Amongst the macro-areas of research emerged from our survey, micro- and macro-programming approaches to IoT system development are the ones concerned the most with the actuation facet, in the sense of programming devices to accomplish tasks. Whereas both SP and CEP are mostly developed with data collection and processing, respectively, in mind and cannot be used as general programming languages executed on-board any device, the programming approaches in this latter broad category are specifically conceived to close this gap. Being general purpose, these approaches may be used for data collection and processing too, but doing so is likely "reinventing the wheel" as SP and CEP are readily available.

2.6 Conclusion and Future Work

From the survey just presented, a few considerations can be made to give a bird-eye view of the whole research landscape about IoT programming, crossing borders across the macro-areas just presented.

The most obvious consideration is that "programming IoT" may actually translate to quite a different activity depending on the specific goals of the application to be built. On the one hand, IoT applications especially interested in monitoring a given system, for example, to provide a dashboard to human supervisors, are usually built by taking off-the-shelf SP or CEP platforms (depending on the sophistication of the required data processing). In this case, "programming" may actually translate to configuring the logical DAG for data processing, that is, deciding which operators to apply to what data and how to put operators in a pipeline, or defining patterns and rules for triggering computations upon detection of a given pattern of events. On the other hand, applications may also be interested in closing the feedback loop between the physical (the things) and the digital (the processing) by sending commands back to the (actuator) devices. In this case, that could regard a swarm robotics scenario, for instance, "programming" usually translates more directly to the traditional programming experience, using mainstream languages (for micro-programming approaches) or Domain-Specific Languages (more common for macro-programming approaches) that are anyway general purpose and often times also closely resemble a mainstream language. This is to say that the "programming experience" may be quite different depending on the kind of IoT systems one wants to develop.

A huge impact on this programming experience is also given by fragmentation of abstractions across the research landscape. In the SP and CEP areas, the abstractions are somewhat similar—cf., the concepts of "stream" (of events or data), operator, composition, pattern, etc. Similarly, in the micro-programming area, abstractions are often the classical ones from mainstream programming (i.e., objects, functions, threads). Conversely, in the realm of macro-programming, essentially any approach has its own peculiar set of abstractions (e.g., fields in aggregate computing, collective tasks in SmartSociety, collective data flows in DDFlow, etc.). Mappings can be devised out, but integration of the different approaches in a single software project nevertheless suffers, as developers need to master different paradigms.

However, such an abundance of approaches also has upsides: Since each programming approach is conceived with a specific set of functional and nonfunctional desiderata, each approach ends up being complementary to others.

Nevertheless, to be fully harnessed, such complementarity must be exploited in an integrated framework, not by patching together different approaches on an ad hoc basis.

References

1. Node-Red. https://nodered.org/. [Online; accessed 19-January-2022]
2. Abadi, D.J., Ahmad, Y., Balazinska, M., Çetintemel, U., Cherniack, M., Hwang, J.-H., Lindner, W., Maskey, A., Rasin, A., Ryvkina, E., Tatbul, N., Xing, Y., Zdonik, S.B.: The design of the borealis stream processing engine. In: Second Biennial Conference on Innovative Data Systems Research, CIDR 2005, Asilomar, CA, USA, January 4–7, 2005, Online Proceedings, pp. 277–289 (2005). www.cidrdb.org
3. Abadi, D.J., Carney, D., Çetintemel, U., Cherniack, M., Convey, C., Erwin, C., Galvez, E.F., Hatoun, M., Maskey, A., Rasin, A., Singer, A., Stonebraker, M., Tatbul, N., Xing, Y., Yan, R., Zdonik, S.B.: Aurora: A data stream management system. In: Halevy, A.Y., Ives, Z.G., Doan, A.H. (eds.) Proceedings of the 2003 ACM SIGMOD International Conference on Management of Data, San Diego, California, USA, June 9–12, 2003, p. 666. ACM, New York (2003)
4. Adi, A., Etzion, O.: Amit—the situation manager. VLDB J. **13**(2), 177–203 (2004)
5. Akbar, A., Carrez, F., Moessner, K., Sancho, J., Rico, J.: Context-aware stream processing for distributed IoT applications. In: 2nd IEEE World Forum on Internet of Things, WF-IoT 2015, Milan, Italy, December 14–16, 2015, pp. 663–668. IEEE Computer Society, New York (2015)
6. Alevizos, E., Skarlatidis, A., Artikis, A., Paliouras, G.: Probabilistic complex event recognition: A survey. ACM Comput. Surv. **50**(5), 71:1–71:31 (2017)

7. Ali, M., Chandramouli, B., Goldstein, J., Schindlauer, R.: The extensibility framework in Microsoft streaminsight. In: 2011 IEEE 27th International Conference on Data Engineering, pp. 1242–1253. IEEE, New York (2011)
8. Angsuchotmetee, C., Chbeir, R.: A survey on complex event definition languages in multimedia sensor networks. In: Chbeir, R., Agrawal, R., Biskri, I. (eds.) Proceedings of the 8th International Conference on Management of Digital EcoSystems, MEDES 2016, Biarritz, France, November 1–4, 2016, pp. 99–108. ACM, New York (2016)
9. Arasu, A., Babcock, B., Babu, S., Datar, M., Ito, K., Motwani, R., Nishizawa, I., Srivastava, U., Thomas, D., Varma, R., Widom, J.: STREAM: the Stanford stream data manager. IEEE Data Eng. Bull. **26**(1), 19–26 (2003)
10. Artikis, A., Skarlatidis, A., Portet, F., Paliouras, G.: Logic-based event recognition. Knowl. Eng. Rev. **27**(4), 469–506 (2012)
11. Azzara, A., Alessandrelli, D., Bocchino, S., Petracca, M., Pagano, P.: PyOT, a macroprogramming framework for the internet of things. In: Proceedings of the 9th IEEE International Symposium on Industrial Embedded Systems, SIES 2014, Pisa, Italy, June 18–20, 2014, pp. 96–103. IEEE, New York (2014)
12. Babcock, B., Babu, S., Datar, M., Motwani, R., Widom, J.: Models and issues in data stream systems. In: Popa, L., Abiteboul, S., Kolaitis, P.G., (eds.) Proceedings of the Twenty-first ACM SIGACT-SIGMOD-SIGART Symposium on Principles of Database Systems, June 3–5, Madison, Wisconsin, USA, pp. 1–16. ACM, New York (2002)
13. Babun, L., Denney, K., Berkay, Z. Celik, McDaniel, P.D., Selcuk Uluagac, A.: A Survey on IoT Platforms: Communication, Security, and Privacy Perspectives. 192:108040
14. Bak, N., Chang, B.-M., Choi, K.: Smart block: a visual block language and its programming environment for IoT. J. Comput. Lang. **60**, 100999 (2020)
15. Beal, J., Dulman, S., Usbeck, K., Viroli, M., Correll, N.: Organizing the aggregate: languages for spatial computing. CoRR, abs/1202.5509 (2012)
16. Beal, J., Pianini, D., Viroli, M.: Aggregate programming for the internet of things. Computer **48**(9), 22–30 (2015)
17. Belson, B., Holdsworth, J., Xiang, W., Philippa, B.: A survey of asynchronous programming using coroutines in the Internet of things and embedded systems. ACM Trans. Embed. Comput. Syst. **18**(3), 21:1–21:21 (2019)
18. Boissier, O., Bordini, R.H., Hubner, J., Ricci, A.: Multi-Agent Oriented Programming: Programming Multi-Agent Systems Using JaCaMo. MIT Press, New York (2020)
19. Carbone, P., Katsifodimos, A., Ewen, S., Markl, V., Haridi, S., Tzoumas, K.: Apache flink: stream and batch processing in a single engine. Bull. IEEE Comput. Soc. Tech. Committee Data Eng. **36**(4), 28–38 (2015)
20. Casadei, R.: Macroprogramming: Concepts, state of the art, and opportunities of macroscopic behaviour modelling. ACM Comput. Surv. **55**(13s), 1–37 (2023)
21. Casadei, R.: Artificial collective intelligence engineering: a survey of concepts and perspectives. Artificial Life **29**(4), 433–467 (2023). 10.1162/artl_a_00408
22. Chandrasekaran, S., Cooper, O., Deshpande, A., Franklin, M.J., Hellerstein, J.M., Hong, W., Krishnamurthy, S., Madden, S., Reiss, F., Shah, M.A.: TelegraphCQ: continuous dataflow processing. In: Halevy, A.Y., Ives, Z.G., Doan, A.H. (eds.) Proceedings of the 2003 ACM SIGMOD International Conference on Management of Data, San Diego, California, USA, June 9–12, 2003, p. 668. ACM, New York (2003)
23. Chaqfeh, M.A., Mohamed, N.: Challenges in middleware solutions for the internet of things. In: 2012 International Conference on Collaboration Technologies and Systems (CTS), p. 21–26 (2012)
24. Choochotkaew, S., Yamaguchi, H., Higashino, T., Shibuya, M., Hasegawa, T.: Edgecep: fully-distributed complex event processing on IoT edges. In: 13th International Conference on Distributed Computing in Sensor Systems, DCOSS 2017, Ottawa, ON, Canada, June 5–7, 2017, pp. 121–129. IEEE, New York (2017)
25. Corno, F., De Russis, L., Sáenz, J.P.: On the challenges novice programmers experience in developing IoT systems: A survey. J. Syst. Softw. **157**, 110389 (2019)

26. Cugola, G., Margara, A.: RACED: an adaptive middleware for complex event detection. In: Grace, P., Eliassen, F. (eds.) Proceedings of the 8th Workshop on Adaptive and Reflective Middleware, ARM 2009, held at the ACM/IFIP/USENIX International Middleware Conference, December 1, 2009, Urbana Champaign, IL, USA, p. 5. ACM, New York (2009).
27. Cugola, G., Margara, A.: Complex event processing with T-REX. J. Syst. Softw. **85**(8), 1709–1728 (2012)
28. Cugola, G., Margara, A.: Processing flows of information: From data stream to complex event processing. ACM Comput. Surv. **44**(3), 15:1–15:62 (2012)
29. Cugola, G., Margara, A.: The Complex Event Processing Paradigm, pp. 113–133. Springer International Publishing, Cham (2015)
30. Dayarathna, M., Perera, S.: Recent advancements in event processing. ACM Comput. Surv. **51**(2), 1–36 (2018)
31. Demers, A.J., Gehrke, J., Panda, B., Riedewald, M., Sharma, V., White, W.M.: Cayuga: A general purpose event monitoring system. In: Third Biennial Conference on Innovative Data Systems Research, CIDR 2007, Asilomar, CA, USA, January 7–10, 2007, Online Proceedings, pp. 412–422 (2007). www.cidrdb.org
32. Dias, J.P., Restivo, A., Ferreira, H.S.: Designing and constructing Internet-of-Things systems: an overview of the ecosystem. Internet of Things **19**, 100529 (2022)
33. Dias de Assunção, M., da Silva Veith, A., Buyya, R.: Distributed data stream processing and edge computing: A survey on resource elasticity and future directions. J. Netw. Comput. Appl. **103**(2017), 1–17 (2018)
34. Doukas, C., Antonelli, F.: Compose: building smart & context-aware mobile applications utilizing IoT technologies. In: Global Information Infrastructure Symposium-GIIS 2013, pp. 1–6. IEEE, New York (2013)
35. Eckert, M., Bry, F., Brodt, S., Poppe, O., Hausmann, S.: A CEP Babelfish: Languages for complex event processing and querying surveyed. In: Reasoning in Event-Based Distributed Systems, pp. 47–70. Springer, Berlin (2011)
36. Fortino, G., Guerrieri, A., Savaglio, C., Spezzano, G.: A review of internet of things platforms through the IoT-A reference architecture. In: Camacho, D., Rosaci, D., Sarné, G.M.L., Versaci, M., (eds.) Intelligent Distributed Computing XIV, 14th International Symposium on Intelligent Distributed Computing, IDC 2021, Virtual Event, 16–18 September 2021. Studies in Computational Intelligence, vol. 1026, pp. 25–34. Springer, Berlin (2021)
37. Fragkoulis, M., Carbone, P., Kalavri, V., Katsifodimos, A.: A survey on the evolution of stream processing systems. CoRR, abs/2008.00842 (2020)
38. Giatrakos, N., Alevizos, E., Artikis, A., Deligiannakis, A., Garofalakis, M.N.: Complex event recognition in the big data era: a survey. VLDB J. **29**(1), 313–352 (2020)
39. Giatrakos, N., Kougioumtzi, E., Kontaxakis, A., Deligiannakis, A., Kotidis, Y.: Easyflinkcep: Big event data analytics for everyone. In: Demartini, G., Zuccon, G., Culpepper, J.S., Huang, Z., Tong, H. (eds.) CIKM '21: The 30th ACM International Conference on Information and Knowledge Management, Virtual Event, Queensland, Australia, November 1–5, 2021, pp. 3029–3033. ACM, New York (2021)
40. TIBCO BusinessEventsUser's Guide. Tibco® business events. In: Software Release, p. 1 (2005)
41. Gyllstrom, D., Wu, E., Chae, H.-J., Diao, Y., Stahlberg, P., Anderson, G.: SASE: complex event processing over streams (demo). In: Third Biennial Conference on Innovative Data Systems Research, CIDR 2007, Asilomar, CA, USA, January 7–10, 2007, Online Proceedings, pp. 407–411 (2007). www.cidrdb.org
42. Hillston, J., Pitt, J., Wirsing, M., Zambonelli, F.: Collective adaptive systems: qualitative and quantitative modelling and analysis (Dagstuhl seminar 14512). Dagstuhl Reports **4**(12), 68–113 (2014)
43. EsperTech Inc. ESPER (2021)
44. Red Hat Inc. Open CEP (2021)
45. STRATIO BIG DATA Inc. Decision CEP Engine (2021)

46. Isah, H., Abughofa, T., Mahfuz, S., Ajerla, D., Zulkernine, F.H., Khan, S.: A survey of distributed data stream processing frameworks. IEEE Access **7**, 154300–154316 (2019)
47. Júnior, I.G.S., Santana, T.S., Bulcão-Neto, R.F., Porter, B.: The state of the art of macroprogramming in IoT: An update. J. Internet Serv. Appl. **13**(1), 54–65 (2022)
48. Kaed, C.E., Khan, I., Van Den Berg, A., Hossayni, H., Saint-Marcel, C.: SRE: Semantic rules engine for the industrial Internet-of-Things gateways. IEEE Trans. Industr. Inform. **14**(2), 715–724 (2018)
49. Khoussainova, N., Balazinska, M., Suciu, D.: Probabilistic event extraction from RFID data. In: Alonso, G., Blakeley, J.A., Chen, A.L.P. (eds.) Proceedings of the 24th International Conference on Data Engineering, ICDE 2008, April 7–12, 2008, Cancún, Mexico, pp. 1480–1482. IEEE Computer Society, New York (2008)
50. Kolchinsky, I., Schuster, A.: Real-time multi-pattern detection over event streams. In: Boncz, P.A., Manegold, S., Ailamaki, A., Deshpande, A., Kraska, T. (eds.) Proceedings of the 2019 International Conference on Management of Data, SIGMOD Conference 2019, Amsterdam, The Netherlands, June 30–July 5, 2019, pp. 589–606. ACM, New York (2019)
51. Kowalski, R.A., Sergot, M.J.: A logic-based calculus of events. New Gener. Comput. **4**(1), 67–95 (1986)
52. Kreps, J., Narkhede, N., Rao, J., et al.: Kafka: a distributed messaging system for log processing. In: Proceedings of the NetDB, vol. 11, pp. 1–7 (2011)
53. Krishnamurthy, J., Maheswaran, M.: Chapter 5—programming frameworks for internet of things. In: Buyya, R., Vahid Dastjerdi, A. (eds.) Internet of Things, pp. 79–102. Morgan Kaufmann, Burlington (2016)
54. Kuhail, M.A., Farooq, S., Hammad, R., Bahja, M.: Characterizing visual programming approaches for end-user developers: a systematic review. IEEE Access **9**, 14181–14202 (2021)
55. Kulkarni, S., Bhagat, N., Fu, M., Kedigehalli, V., Kellogg, C., Mittal, S., Patel, J.M., Ramasamy, K., Taneja, S.: Twitter heron: stream processing at scale. In: Proceedings of the 2015 ACM SIGMOD International Conference on Management of Data, pp. 239–250 (2015)
56. Li, B., Dong, W.: EdgeProg: Edge-centric programming for IoT applications. In: 2020 IEEE 40th International Conference on Distributed Computing Systems (ICDCS), pp. 212–222. IEEE, New York (2020)
57. Li, G., Jacobsen, H.-A.: Composite subscriptions in content-based publish/subscribe systems. In: Alonso, G., (ed.) Middleware 2005, ACM/IFIP/USENIX, 6th International Middleware Conference, Grenoble, France, November 28–December 2, 2005, Proceedings. Lecture Notes in Computer Science, vol. 3790, pp. 249–269. Springer, Berlin (2005)
58. Lohstroh, M., Kim, H., Eidson, J.C., Jerad, C., Osyk, B., Lee, E.A.: On enabling technologies for the internet of important things. IEEE Access **7**, 27244–27256 (2019)
59. Luckham, D.: The Power of Events, vol. 204. Addison-Wesley Reading, New York (2002)
60. Luckham, D.: What's the difference between ESP and CEP (2006). https://complexevents.com/2020/06/15/whats-the-difference-between-esp-and-cep-2/
61. Luckham, D.: Causality in Event Stream Analytics (2020). https://complexevents.com/2020/08/27/causality-in-event-stream-analytics/
62. Luckham, D.C.: Event Processing for Business: Organizing the Real-time Enterprise. Wiley, New York (2011)
63. Madden, S., Szewczyk, R., Franklin, M.J., Culler, D.E.: Supporting aggregate queries over ad-hoc wireless sensor networks. In: 4th IEEE Workshop on Mobile Computing Systems and Applications (WMCSA 2002), 20–21 June 2002, Callicoon, NY, USA, pp. 49–58. IEEE Computer Society, New York (2002)
64. Mahmood, N., Pasha, M.K., Pasha, K.: Survey of applications of complex event processing (CEP) in health domain. Sukkur IBA J. Comput. Math. Sci. **1**(2), 88–94 (2017)
65. Mizzi, A., Ellul, J., Pace, G.: D'Artagnan: An embedded DSL framework for distributed embedded systems. In: Proceedings of the Real World Domain Specific Languages Workshop 2018, pp. 1–9 (2018)

66. Namiot, D., Sneps-Sneppe, M.: On IoT programming. Int. J. Open Inf. Technol. **2**(10), 25–28 (2014)
67. Nasiri, H., Nasehi, S., Goudarzi, M.: Evaluation of distributed stream processing frameworks for IoT applications in smart cities. J. Big Data **6**, 52 (2019)
68. Newton, R., Welsh, M.: Region streams: functional macroprogramming for sensor networks. In: Labrinidis, A., Madden, S., (eds.) Proceedings of the 1st Workshop on Data Management for Sensor Networks, in Conjunction with VLDB, DMSN 2004, Toronto, Canada, August 30, 2004. ACM International Conference Proceeding Series, vol. 72, pp. 78–87. ACM, New York (2004)
69. Ngu, A.H., Gutierrez, M., Metsis, V., Nepal, S., Sheng, Q.Z.: IoT Middleware: a survey on issues and enabling technologies. IEEE Internet Things J. **4**(1), 1–20 (2017)
70. Nguyen, X.T., Tran, H.T., Baraki, H., Geihs, K.: FRASAD: A framework for model-driven IoT application development. In: 2015 IEEE 2nd World Forum on Internet of Things (WF-IoT), pp. 387–392. IEEE, New York (2015)
71. Noghabi, S.A., Paramasivam, K., Pan, Y., Ramesh, N., Bringhurst, J., Gupta, I., Campbell, R.H.: Samza: stateful scalable stream processing at LinkedIn. Proc. VLDB Endow. **10**(12), 1634–1645 (2017)
72. Noor, J., Tseng, H.-Y., Garcia, L., Srivastava, M.B.: DDFlow: visualized declarative programming for heterogeneous IoT networks. In: Landsiedel, O., Nahrstedt, K. (eds.) Proceedings of the International Conference on Internet of Things Design and Implementation, IoTDI 2019, Montreal, QC, Canada, April 15–18, 2019, pp. 172–177. ACM, New York (2019)
73. Oracle. Oracle CEP (2021)
74. Patel, P., Cassou, D.: Enabling high-level application development for the internet of things. J. Syst. Softw. **103**, 62–84 (2015)
75. Perera, C., Liu, C.H., Jayawardena, S., Chen, M.: A survey on internet of things from industrial market perspective. IEEE Access **2**, 1660–1679 (2014)
76. Power, A., Kotonya, G.: Bobocep: Distributed complex event processing for resilient fault-tolerance support in IoT. In: 6th IEEE International Conference on Big Data Computing Service and Applications, BigDataService 2020, Oxford, UK, August 3–6, 2020, pp. 109–112. IEEE, New York (2020)
77. Qi, J., Yang, P., Waraich, A., Deng, Z., Zhao, Y., Yang, Y.: Examining sensor-based physical activity recognition and monitoring for healthcare using internet of things: a systematic review. J. Biomed. Inform. **87**, 138–153 (2018)
78. Raglin, A., Venkateswaran, A., Liu, H.: Abductive causal reasoning for Internet of Things. In: 2019 IEEE SmartWorld, Ubiquitous Intelligence Computing, Advanced Trusted Computing, Scalable Computing Communications, Cloud Big Data Computing, Internet of People and Smart City Innovation (SmartWorld/SCALCOM/UIC/ATC/CBDCom/IOP/SCI), pp. 1866–1869 (2019)
79. Ray, P.P.: A survey on visual programming languages in internet of things. Sci. Program. **2017**, 1231430:1–1231430:6 (2017)
80. Rayns, C., Ritchie, A., Balakrishnan, S., Clark, D., Coxhead, P., Donnelly, D., Ghosh, K., Millwood, D., Peulvast, N., Vanstone, I., et al.: Patterns: Integrating WebSphere ILOG JRules with IBM Software. IBM Redbooks, New York (2011)
81. Razzaque, M.A., Milojevic-Jevric, M., Palade, A., Clarke, S.: Middleware for internet of things: A survey. IEEE Internet Things J. **3**(1), 70–95 (2016)
82. Röger, H., Mayer, R.: A comprehensive survey on parallelization and elasticity in stream processing. ACM Comput. Surv. **52**(2), 1–37 (2019)
83. Sahal, R., Breslin, J.G., Ali, M.I.: Big data and stream processing platforms for Industry 4.0 requirements mapping for a predictive maintenance use case. J. Manuf. Syst. **54**, 138–151 (2020)
84. Salman, A.J., Al-Jawad, M., Tameemi, W.A.: Domain-specific languages for IoT: challenges and opportunities. In: IOP Conference Series: Materials Science and Engineering, vol. 1067(1), p. 012133. IOP Publishing, New York (2021)
85. Ltd. Samsung Electronics Co. Spark CEP (2021)

86. Savaglio, C., Fortino, G., Ganzha, M., Paprzycki, M., Bădică, C., Ivanović, M.: Agent-Based Computing in the Internet of Things: A Survey, pp. 307–320. Springer International Publishing, Cham (2018)
87. Savaglio, C., Ganzha, M., Paprzycki, M., Badica, C., Ivanovic, M., Fortino, G.: Agent-based internet of things: State-of-the-art and research challenges. Future Gener. Comput. Syst. **102**, 1038–1053 (2020)
88. Scekic, O., Schiavinotto, T., Videnov, S., Rovatsos, M., Truong, H.L., Miorandi, D., Dustdar, S.: A programming model for hybrid collaborative adaptive systems. IEEE Trans. Emerg. Top. Comput. **8**(1), 6–19 (2020)
89. Schultz-Møller, N.P., Migliavacca, M., Pietzuch, P.R.: Distributed complex event processing with query rewriting. In: Gokhale, A.S., Schmidt, D.C. (eds.) Proceedings of the Third ACM International Conference on Distributed Event-Based Systems, DEBS 2009, Nashville, Tennessee, USA, July 6–9, 2009. ACM, New York (2009)
90. Singh, K.J., Kapoor, D.S.: Create your own Internet of Things: a survey of IoT platforms. IEEE Consum. Electron. Mag. **6**(2), 57–68 (2017)
91. Skarbovsky, I.: IBM Proactive Technology Online (PROTON) (2021)
92. Soulier, P., Li, D., Williams, J.R.: A survey of language-based approaches to cyber-physical and embedded system development. Tsinghua Sci. Technol. **20**(2), 130–141 (2015)
93. Stonebraker, M., Çetintemel, U., Zdonik, S.: The 8 requirements of real-time stream processing. ACM Sigmod Record **34**(4), 42–47 (2005)
94. Suhothayan, S., Gajasinghe, K., Narangoda, I.L., Chaturanga, S., Perera, S., Nanayakkara, V.: Siddhi: a second look at complex event processing architectures. In: Dooley, R., Fiore, S., Green, M.L., Kiddle, C., Marru, S., Pierce, M.E., Thomas, M., Wilkins-Diehr, N. (eds.) Proceedings of the 2011 ACM SC Workshop on Gateway Computing Environments, GCE 2011, Seattle, WA, USA, November 18, 2011, pp. 43–50. ACM, New York (2011)
95. Tantalaki, N., Souravlas, S., Roumeliotis, M.: A review on big data real-time stream processing and its scheduling techniques. *Int. J. Parallel Emergent Distrib. Syst.* **35**(5), 571–601 (2019)
96. The JBoss Drools team. Drools Fusion User Guide (2021)
97. Thakkar, R., Bhavsar, M.: Achieving multilevel elasticity for distributed stream processing systems in the cloud environment: A review and conceptual framework. In: Proceedings of the 2022 Fourteenth International Conference on Contemporary Computing, IC3-2022, Noida, India, August 4–6, 2022, pp. 81–90. ACM, New York
98. Toshniwal, A., Taneja, S., Shukla, A., Ramasamy, K., Patel, J.M., Kulkarni, S., Jackson, J., Gade, K., Fu, M., Donham, J., et al.: Storm@ twitter. In: Proceedings of the 2014 ACM SIGMOD International Conference on Management of Data, pp. 147–156 (2014)
99. Udoh, I.S., Kotonya, G.: Developing IoT applications: challenges and frameworks. IET Cyber-Phys. Syst.: Theory Appl. **3**(2), 65–72 (2018)
100. Viroli, M., Beal, J., Damiani, F., Audrito, G., Casadei, R., Pianini, D.: From distributed coordination to field calculus and aggregate computing. J. Log. Algebraic Methods Program. **109**, 100486 (2019)
101. Wan, K., Hughes, D., Man, K.L., Krilavicius, T.: Composition challenges and approaches for cyber physical systems. In: Proceedings of the 1st IEEE International Conference on Networked Embedded Systems for Enterprise Applications, NESEA 2010, November 25–26, 2010, Suzhou, China, pp. 1–7. IEEE Computer Society, New York (2010)
102. Wenger, R., Zhu, X., Krishnamurthy, J., Maheswaran, M.: A programming language and system for heterogeneous cloud of things. In: 2nd IEEE International Conference on Collaboration and Internet Computing, CIC 2016, Pittsburgh, PA, USA, November 1–3, 2016, pp. 169–177. IEEE Computer Society, New York (2016)
103. Wu, E., Diao, Y., Rizvi, S.: High-performance complex event processing over streams. In: Chaudhuri, S., Hristidis, V., Polyzotis, N. (eds.) Proceedings of the ACM SIGMOD International Conference on Management of Data, Chicago, Illinois, USA, June 27–29, 2006, pp. 407–418. ACM, New York (2006)

104. Yasumoto, K., Yamaguchi, H., Shigeno, H.: Survey of real-time processing technologies of IoT data streams. J. Inf. Process. **24**(2), 195–202 (2016)
105. Zagrouba, R., Kardi, A.: Comparative study of energy efficient routing techniques in wireless sensor networks. Information **12**(1), 42 (2021)
106. Zaharia, M., Chowdhury, M., Franklin, M.J., Shenker, S., Stoica, I., et al.: Spark: cluster computing with working sets. HotCloud **10**(10–10), 95 (2010)
107. Zeuch, S., Zacharatou, E.T., Zhang, S., Chatziliadis, X., Chaudhary, A., Monte, B.D., Giouroukis, D., Grulich, P.M., Ziehn, A., Markl, V.: NebulaStream: complex analytics beyond the cloud. Open J. Internet Things **6**(1), 66–81 (2020)
108. Ziehn, A.: Complex event processing for the internet of things. In: Abedjan, Z., Hose, K. (eds.) Proceedings of the VLDB 2020 PhD Workshop co-located with the 46th International Conference on Very Large Databases (VLDB 2020), ONLINE, August 31–September 4, 2020. CEUR Workshop Proceedings, vol. 2652. CEUR-WS.org (2020)

Part II
Fluidware Technologies

Chapter 3
Middleware Architectures for Fluid Computing

Roberto Casadei, Nicolas Farabegoli, Giancarlo Fortino, Claudio Savaglio, and Mirko Viroli

3.1 Introduction

Device, Edge/Fog, and Cloud are the typical hierarchical architecture featuring the majority of current IoT systems, where tasks are statically and a priori situated per layer (e.g., mere data forwarding or simple aggregation at IoT layer, more complex data processing toward the Cloud) regardless of the actual resource availability of the devices. However, such a siloed task assignment collides with the opportunistic essence of the IoT domain (where the QoS requirements should dynamically drive the identification of a target device for a given task) and with the rapid advancements in miniaturization and chips construction, which enabled even small and/or mobile IoT devices like smartphone or Raspberry for example, to potentially behave as an Edge Server, being equipped with performing hardware like multiprocessors and GPU. Moreover, the distribution of the computing in such a strongly layered architecture is proved as effective exclusively intralayer, thus limiting the development of IoT services whose operations should span simultaneously interlayer. Conversely, the fundamental concept underlying Fluidware is to operate within a continuously changing infrastructure that encompasses a spectrum of IoT fabric (i.e., IoT, Edge, and Cloud devices) and resources: This allows for swift adaptation to the internal dynamics of the environment and enables the coordinated execution of distributed activities across devices possessing different levels of computational capability. Therefore, the Fluidware abstractions introduced in the previous chapters (Funnel, Field, Fabric, etc.) need to be grounded on a solid architecture and specific

R. Casadei · N. Farebegoli · M. Viroli
Università di Bologna, Cesena, FC, Italy
e-mail: roby.casadei@unibo.it; nicolas.farabegoli@unibo.it; mirko.viroli@unibo.it

G. Fortino · C. Savaglio (✉)
Università della Calabria, Rende, CS, Italy
e-mail: giancarlo.fortino@unical.it; csavaglio@dimes.unical.it

F. Zambonelli et al. (eds.), *Fluidware*, Internet of Things,
https://doi.org/10.1007/978-3-031-62146-8_3

components specifically outlined to mirror and to support the features and the desiderata of our project. To this end, in this chapter first we briefly analyze main limitations of typical centralized architectures in Sect. 3.2, which led to the "IoT-Edge-Cloud" continuum, and then we introduce our solution in Sect. 3.3, followed by an in-depth about its key element, namely the Fluidware middleware in Sect. 3.4. A particular instantiation of the Fluidware middleware pivoted around the "pulverization" computing paradigm is presented in Sect. 3.5, while final remarks conclude the chapter.

3.2 Motivations: From Fully Centralized to Fully Distributed "Fluid" Architectures

At the state of the art, most of the IoT services still mirror conventional web services (being based on the web, they are often referred to as Web of Things services [17]) and expose a two-level, service-oriented architecture (SOA) [9], like the one reported in Fig. 3.1: In such a setting, service provision is implemented in a centralized way, by having all data from sensors be redirected to some central cloud, where it can be analyzed and—depending on the services to be realized—commands for the actuators can be eventually issued. Typical examples are the Smart Home Assistants: They rely on cloud-based IoT platforms [1] where thermostats, lights, security systems, and appliances forward data for being processed, analyzed, and used to trigger actions or provide insights to the users. In general, these IoT services based on **Cloud Computing** are easily recognizable:

- Device level: Also identified as "perception level," it encompasses all the devices (more or less "smart," i.e., RFDI tags and readers, environmental sensors, wearable devices, video cameras, etc.) able to directly interact with physical

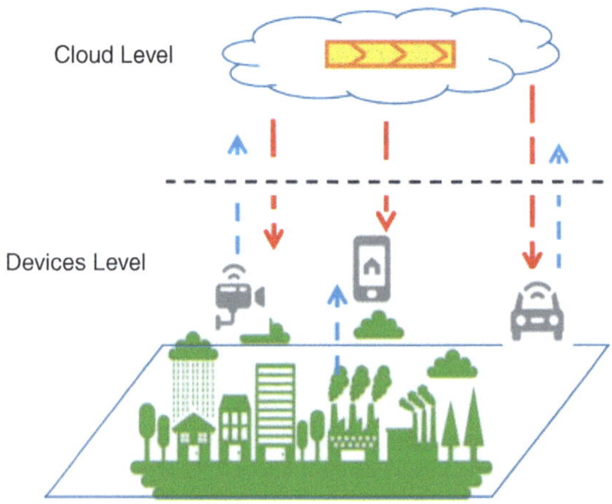

Fig. 3.1 Two conventional two-level architecture based on sense-and-forward approach

phenomena (e.g., temperature or humidity) or real-world events (e.g., window opened, button pressed) and, hence, to communicate with the Cloud to off-load their data, raw or simply refined (e.g., aggregated temperature values on an hourly window, number of people passing their badge).
- Cloud level: It aims to gather all the available data (live, historical, etc.) in huge repositories hosted by powerful remote servers for feeding resource-demanding elaboration processes and, possibly, sending actuation commands back to the Device level (e.g., thermal comfort).

Being adopted by many Cloud IoT platforms such as Microsoft Azure, Amazon AWS, etc., this simple "sense-and-forward" approach originally raised (early 2000s) to address the resources limitation typical of Wireless Sensors and Actuators Networks (WSANs) [15]: These were the only resources of the Device level, strictly limited in terms of storage/processing capabilities and, hence, forced to periodically off-load (almost raw) data. Moreover, the confluence of sensors data into a unique point helped in the development of high-performance AI models, which were trained over huge dataset. However, over the years, the exploding number of heterogeneous IoT devices (referred to as IoT fabric), (Big) data, and services did such a Cloud-dependent approach collide with their inherent real-time nature and their non-negligible requirements in terms of privacy, reliability, adaptivity, and scalability. For example, in the e-Health and Ambient Assisted Living domains [2], sensitive patients' data need to be analyzed "in situ" and immediately flow by the proper actuator if we aim to prevent or assess potentially dangerous health conditions or simply dangerous behaviors and, hence, to take immediate actions. In such a context (or in similar ones, like Smart Transportation or Virtual Reality), the natural nondeterminism and vulnerability of the network made the Cloud, even the most powerful, as a bottleneck for the provision of many key IoT services. Therefore, in the last two decades, distributed architectures according to the **Fog Computing** (from 2012) and **Edge Computing** (from 2016) have arisen with the intent of decentralizing and bringing the processing of IoT data closer to the event sources [10, 12]. In particular, a well-defined set of activities characterizes each level of their architectures so that system administrators deploy each service at the most appropriate location directly at design time:

- Device/Edge level: At the front-end, IoT devices and Edge nodes still perform not only sensory and actuation tasks but also additional, lightweight activities like data compression, filtering, caching, and human interfacing (e.g., through smartphones) or local communication.
- Fog level: At the near-end, Fog nodes are purposely deputed at local ML inference tasks, orchestration and coordination of services and resources, provided with relevant resources and stable Internet connectivity.
- Cloud level: At the far-end, cloud servers are still involved in massive parallel data processing, knowledge discovery, persistent storage, ML models training, etc. and in data exploration (e.g., dashboarding).

For example, with respect to the aforementioned AAL scenario, it would be necessary: (i) for data processing, a web service stack which realizes the system functionalities; (ii) for sending data, the appropriate communications with sensor and actuator devices connected to a local gateway (e.g., a RaspberryPi or a desktop computer). Cooperation between services may happen in a point-to-point fashion through simple RESTful endpoints or by relying on a publish–subscribe broker. Finally, deployment would be mostly decided at design time: The web service back-ends would be likely deployed on the Cloud or possibly to a sufficient powerful Fog node, and devices connected to the gateway would have their own logic embedded to send data and listen for commands. In such a direction, it happens that many Cloud IoT platforms have started embedding models, technologies, and protocols specifically devoted to connect the Cloud to the Edge and Fog levels, where it emerges servitization paradigms different from SOA: Microservices, for example, can be developed, deployed, and scaled independently, thus improving modularity, flexibility, resilience, and interoperability (but at cost of additional complexity in terms of service orchestration, deployment, and monitoring).

While both Fog and Edge paradigms address the challenges of distributing and outsourcing the computing from the cloud while preserving real-timeness, bandwidth consumption, and data privacy, their architectural designs and deployment models differ mainly for (i) the proximity to the data source, since "fog nodes" are intermediate yet powerful devices located between Edge devices and the cloud servers, and (ii) the network connectivity requirements, since "Edge nodes" can operate in environments with intermittent or limited network connectivity, being able to function autonomously and make decisions locally with negligible data transfer. Moreover, with the great advancements and consequent diffusion of single-board computers (SBCs, e.g., Google Coral Board, Nvidia Jetson Nano, RaspberryPi, and Intel NeuroStick) and flash memories, most of the AI/ML workload and storage tasks initially aimed at fog nodes have efficiently carried out by Edge nodes directly at the real data/event source, the reason why the Fog computing has rapidly lost attractiveness to the advantage of Edge Computing. Finally, more recently (2020), the idea of an architecture with clearly marked and compartmentalized levels has been replaced with the seamless **"IoT-Edge-Cloud" continuum** [16]. This can be visualized as a gradient, where all the aforementioned computing paradigms are integrated and, potentially, interchangeably into an opportunistic manner, accordingly to the specific needs of IoT services and the actual availability or resources. In such an approach, there are no a priori-defined responsibilities or deployment targets, but these are case-by-case established by means of purposed middleware components operating according to specific policies. This provides great flexibility and exactly matches the recent AI trends, heading toward a dynamic distribution of traditional models and techniques (at a different degree of accuracy) across a spectrum of heterogeneous computing resources, toward an actionable intelligence [14], i.e., the so-called Edge Intelligence [4].

3.3 The Fluidware Architecture

To overcome the aforementioned limitations of a compartmentalized two-/three-layered architecture and by adhering to the continuum principles, we outline a fully distributed, super-peer architecture where Fluidware services are supported by IoT devices, Edge nodes, and/or Cloud platforms which are dynamically activated and reconfigured.

Consequently, instead of conventional layered architecture based upon a static task assignment, we outline an organization based on abstraction levels so that developers willing to design an IoT service or application in terms funnels and event streams detached from the physical devices actually producing and consuming them:

- Device abstraction level: encompassing sensor and actuator networks
- Event abstraction level: encompassing streams of data flowing in both directions, i.e., "up-flows" and "down-flows"
- Funnel process abstraction level: encompassing funnel processes which digest contextualized streams of events

Indeed, to promote scalability and flexibility, Fluidware will support three kinds of interactions and coordination (as in Fig. 3.2):

- Direct device-to-device interactions (e.g., for field-based coordination) [6, 11], with funnel processes directly instantiated on sets of devices and with events flowing, aggregating, and redistributing from device to device
- Intra-edge and inter-edge interactions, with funnel processes dynamically allocated on edge computers (i.e., cloudlets or fog computers), to digest streams of events, implement local coordinated services, and possibly connect multiple edges to realize inter-edge coordination

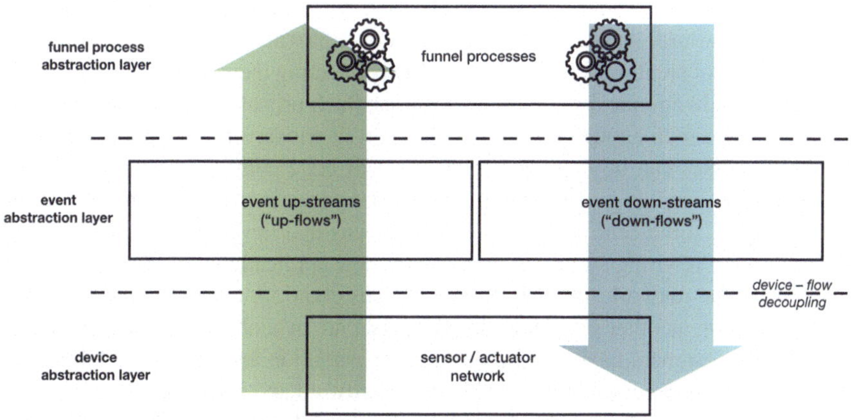

Fig. 3.2 Three-layer architecture with different abstraction degrees

- Cloud-based interactions, for centralized monitoring, coordination, and storage

The outlined architecture based on the continuum concept over-performs conventional 2-/3-layered architectures for many reasons: WoT requires a full network stack to be available on connected devices, which is impractical for the resource-constrained ones usually employed at the Edge. SOA does not cope well with moving services across tiers' boundaries depending on available resources. The 2-/3-tiered network topology model is usually accepted as a static one, where services are allocated to one specific tier at design time, and thus adding a new device to the system usually implies adding a new service to the stack, taking care of connecting the device to the rest of the application, rewiring service composition routing paths, and such. Also, scaling the service requires in most cases a redeployment of the whole application.

The execution of funnel processes along the entire IoT/edge/cloud stack is enabled by the key component of the Fluidware architecture, the middleware, which fully provides utility-driven exploitation of infrastructure, and it is described in the following section.

3.4 The Fluidware Middleware

Middleware, widely used in conventional distributed systems, provides general and specific abstractions (e.g., object computation model, inter-object communication, hardware and software interfaces, data types, discovery service, knowledge management, and physical location) through which IoT devices and systems and their related services can be easily and securely built up and integrated. Therefore, middleware plays a key role for the design and implementation of the whole [8, 13] acting as a bridge between the IoT devices and the services that interact with them.

As anticipated in Chap. 1, funnel processes will be specified in a declarative way independently of their actual allocation and distribution, being managed by the Fluidware middleware. At the implementation and deployment level, the middleware will take care of creating actual processes and connecting them to the flows of actual devices, to realize funnel process specifications. Such processes can be deployed on any IoT device with enough resources to support their execution. Opportunistically, funnel processes can be deployed at the level of edge computers associated with some specific location and having access to all devices in that location [3] or even at the level of some centralized cloud. This means that developers need no longer to bother with rewiring service connections if new applications are deployed, or to reconfigure the network topology if new devices come in, nor to deal with device failures as long as other providing the same data are available. Overall, this means that system designers have not only a higher abstraction level provided but also a higher degree of decoupling between system hardware and software components.

In more detail, we expect the Fluidware middleware, possibly guided by some "allocation" directive, to automatically handle the allocation and possible replication of Funnel processes, their optimized execution, and transparent replication/relocation across the network [18]. Specifically, the activities to develop the Fluidware middleware platform will address five related objectives:

1. At run time, to realize the funnel processes specification, by connecting the event streams to the actual source devices' data and establishing which flows to connect to a given funnel process (e.g., all flows in target area, only flows with certain properties, etc.), and how (e.g., as a message stream, through a publish–subscribe blackboard, or with a tuple space)
2. Define a mapping from the funnels processes specifications into a set of distributed components to be deployed atop the Fluidware platform—such mapping will be based on a model-driven development and will distinguish abstract platform-independent specifications from deployable platform-dependent components
3. To deploy such a funnel process over the platform-dependent components that can be dynamically activated and reconfigured in IoT devices, Edge servers, and/or Cloud platforms as individual process, or as part of a swarm of distributed replicas, or as a worker in a map-reduce-like architecture
4. To dynamically manage the deployed funnels by instantiating local proxies of processes and to relocate them as needed
5. To analyze interoperability and security issues, by defining guidelines for enabling devices to connect to the Fluidware platform and analyzing how a trust-oriented distributed infrastructure for inter-component security can be integrated within it

For instance, a funnel process collecting temperature measures from a given area to chart a heatmap of that area may be actually composed by three subprocesses: one gathering temperatures from sensors, thus directly deployed onboard Edge devices, one aggregating results to average out temperatures depending on sub-areas (e.g., the kitchen, the living room, first floor, second floor, left wing, right wing, etc.), hence deployed in Fog gateways, and the last one charting the heatmap, deployed on the Cloud (as it needs global information and more computational capacity). Such subprocesses, which can be funnel processes themselves, are opportunistically connected to the most suitable event flows (Fig. 3.2) and deployed to the most suitable device (Fig. 3.3). For instance, should the Fluidware middleware detect failure of a sensor, it could replace the connection to the event flow coming from such sensor with another, non-faulty one. Also, should the middleware detect a performance drop in processing at the Edge, it could decide to relocate the correspondent subprocess to an available Fog gateway.

3.4.1 Toward the Fluidware Middleware Implementation

The outlined abstract components of the Fluidware middleware need to be grounded on a solid architecture and a flexible computing paradigm: Indeed, the achievement of the aforementioned objectives demands for applications deployable across the

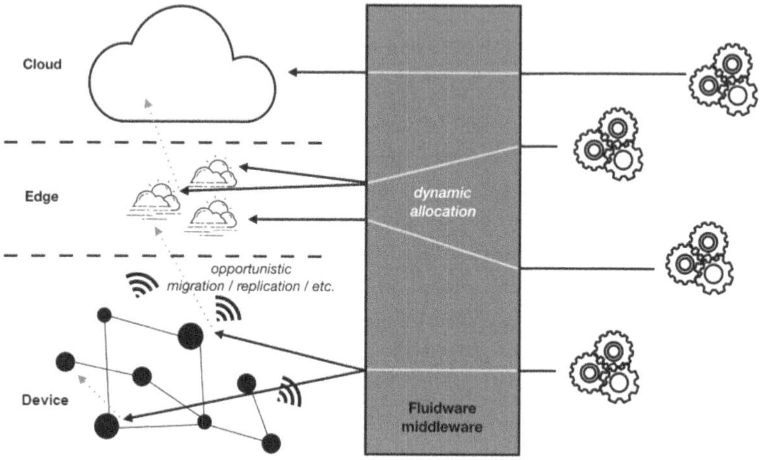

Fig. 3.3 Fluidware middleware deploys funnel processes opportunistically

ICT infrastructure, thus allowing multiple potential deployment configurations for the same application logic with potentially different performance and costs. Therefore, in the following we introduce the "pulverization" paradigm that inherently supports essential features of autonomy, decentralization, and adaptiveness and that allows exploring and assessing the set of potential deployment configurations.

3.5 The *Pulverization* Approach to Distributed Application Partitioning and Deployment

The target infrastructures of IoT applications may vary depending on the application requirements and the available resources. Regarding the infrastructure, we consider the Edge-Cloud continuum, where a multilayer heterogeneous network of devices ranging from cloud servers to small embedded devices represents a valuable solution for the deployment of IoT applications. Such applications can benefit from this kind of infrastructure; however, the effective exploitation of the Edge-Cloud continuum poses new challenges to application designers. Commonly, the design of an application is made having in mind a specific infrastructure, thus making it difficult to exploit the full potential of dynamic and heterogeneous infrastructures.

3.5.1 From Monolith to Component-Based Software Systems

The traditional approach to software system design relies on the modeling of the system as a monolith. In this approach, the different software components and functionalities coexist in the same codebase, and typically the software is distributed and operated as a single deployment unit. Although this approach has evident advantages like *simple design*, *performance* and *security*, and *simple operation*, the monolith nature of this kind of software makes it difficult to evolve the software, resulting in a lower flexibility.

With the emergence of the Cloud, this kind of architecture exhibited several limitations. One of the main advantages of the Cloud is its scalability and flexibility depending on the changing requirements; these characteristics cannot be fully exploited with monolithic applications. The complexity is represented by the fact that if scaling is required, the replication of the entire software system is required, when it is likely that only one part of the system should be subject to replication, resulting in a waste of resource with less effective results.

To better exploit the full potential of Cloud services, The recent software system is designed to be highly *modular*, *extensible*, and *flexible*. With this approach, the software system is broken down into several independent software components that communicate each other via specific API. In this way, each component can be updated, modified, or replicated without affecting the rest of the system. Moreover, this modular nature of software systems promotes dynamic reconfiguration according to the changing requirements of the system such as cost, bandwidth, storage, etc. Additionally, with the subdivision in components, we can demand certain responsibilities to other components, distributing and organizing the complexity in a system.

3.5.2 The **Pulverization** *Approach*

In the context of CAS (Collective Adaptive System), an effective way of decomposing an application in order to achieve flexible deployment is the *pulverization* approach [7]. This approach fosters the decoupling of the application logic from deployment aspects, as it is explained in the following. Our target systems consist of collections of devices of two kinds:

1. *Application-level devices*: These are the entities that we would like to program or control to provide application-level services.
2. *Infrastructure-level devices*: These are computers that are *not* part of the application itself but rather support its *operational needs*.

In the first place, the pulverization approach distinguishes a *logical system* from the *software system* that implements it and the *physical system* that supports the execution of the implementation. By a logical perspective, a system consists of a network of *devices*, equipped with *sensors* and *actuators*, that can interact by exchanging messages with a dynamic subset of other devices known as a device's *neighborhood*. Each device has a *behavior* organized into *rounds*, with each round logically consisting into *sense–compute–interact* steps.

A simple deployment and operational scheme for this logical system could consist in deploying one software component per logical device into the corresponding application-level physical device. However, this has various drawbacks: (i) Resource-constrained devices may not be able to execute the software components. (ii) The exploitation of the infrastructural elements is limited. (iii) The deployment solution space is very small, hence the possibility for optimizations.

Then, the idea of the pulverization approach consists in designing the activity of each logical device in terms of an ensemble of *five components*:

1. *Behavior*
2. *State*
3. *Communication*
4. *Sensors*
5. *Actuators*

With this design, each component can be implemented as a *software component*, also working as a *deployment unit*. All five components can be deployed independently on the available infrastructure without affecting the application logic. In this way, the designer can focus on the business logic of the application, delaying the decision about the deployment infrastructure to the deployment phase.

Figure 3.4 exemplifies the pulverization model and examples of deployments.

Each device, in the pulverization model, performs a well-defined cycle composed of the following steps: *context acquisition* where information about the context (e.g., sensors' state, neighbors' messages) is collected, *computation* where the behavior function is applied against the context information, *coordination* where from the state, coordination messages are sent to neighbor devices, and finally, *actuation* where the actuators are activated according to the device state. The continuous execution of this loop determines the overall behavior of the device.

Once the application has been "pulverized," a mapping must be provided between the pulverized logical system and the hosts that will execute the pulverized components. In this vision, a single logical device usually ends up being executed on multiple hosts.

3.5.2.1 Target Infrastructure and Applications

The pulverization approach tries to tackle the complex problem of deploying CASs in the context of CPS, where a huge number of devices should coordinate with one another to achieve a common goal, and sensing and acting operations are required to interact with the physical environment. Moreover, due to the high dynamic nature of CAS, modern infrastructure such as the Cloud-Edge continuum can provide an effective solution for the deployment of this kind of systems. The intrinsic complexity of the managing of this type of infrastructure makes the pulverization approach and effective solution for hiding it and letting to the platform/middleware its management.

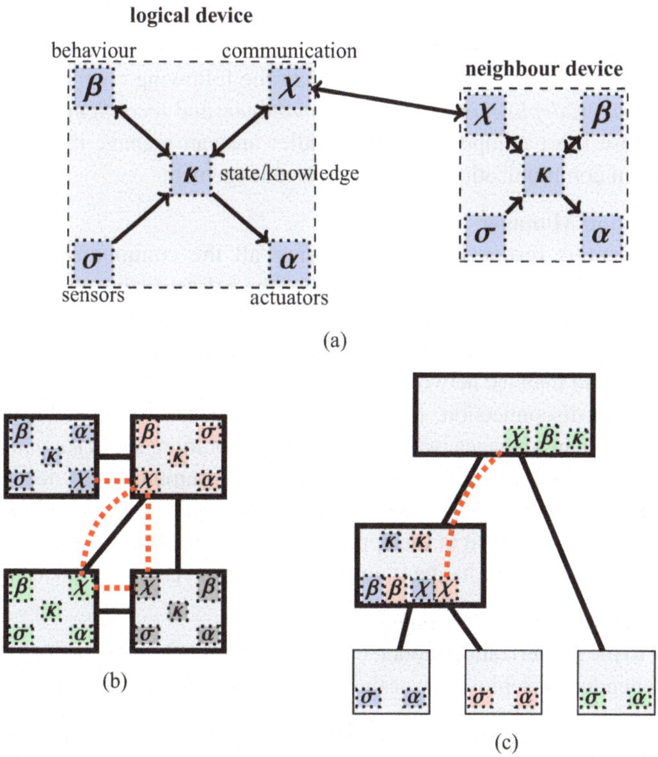

Fig. 3.4 Pulverization model and examples of deployments (figure adapted from [5]). Notation: Solid-border boxes denote physical hosts (bold borders are for thick devices); solid lines denote connections between hosts; dashed-border boxes denote software components; different colors denote (software components of) different logical devices; and red dashed lines denote connections between the software components, i.e., neighboring relationships (not shown for colocated software components). (**a**) A logical device, split into sub-components, and one of its neighbors. (**b**) Peer-to-peer architecture: one-to-one mapping between logical and physical devices, with no offloading. (**c**) IoT hosts can be thin, with some components offloaded at the Edge and Cloud

Relevant scenarios in which the pulverization approach can be applied are *IoT Systems*, *Smart Cities*, *Wearable Computing Systems*, or more generally, in all scenarios characterized by a high level of heterogeneity in terms of computational resources, devices, etc., and cooperation and coordination between the devices are required.

3.5.3 Pulverization *Middleware*

The *pulverization* approach requires the support of a middleware for executing one or more applications. In this section we will provide a possible architecture of a middleware supporting the pulverization approach, the main components and responsibilities, and the expected outcomes of the middleware.

A possible architecture can be composed of the following components: *communication manager*, *deployment unit control manager*, and *reconfiguration manager*. Based on these three components, the middleware can manage different aspects, spanning from communication to component management.

Communication Manager
This component is responsible for managing all the communications with the other middleware instance distributed among the infrastructure of the system. In particular, this component is responsible for determining the physical network topology upon which the communications of the system are implemented. Ideally, this component should manage network-related operation like component to component communication, disconnection, reshape of the net in terms of reachable devices, etc. This component manages two types of communication: *service communication* and *application communication*. The former are communication needed by the middleware to build the net graph and react to network changes and events. The latter are the communication at the application level, specifically the messages that the components need to exchange to implement the pulverization device cycle.

Deployment Unit Control Manager
According to the pulverization model, each component belonging to a logical device can be independently deployed over the hosts that constitute the infrastructure. This component of the middleware is responsible for taking into account the execution (the start and the stop) of a component instance, according to the deployment plan generated by the **Reconfiguration manager**, managing the so-called deployment unit: a set of components that belong to a specific instance of a logical device. To support the execution of multiple systems at the same time in the same infrastructure, this component can manage multiple deployment units belonging to logical devices of different applications. In this way, we can efficiently exploit the underlying infrastructure by maximizing the number of applications that can coexist on a single infrastructure.

Reconfiguration Manager
To exploit the full potential of the underlying infrastructure, we need a mechanism to dynamically adapt the deployment strategy according to the changing requirements or condition of the system. This component represents the way we can achieve the dynamic reconfiguration of the system, relying, ideally, on declarative reconfiguration rule expressing the needs of maintaining certain QoS or the condition of the system. Based on the information produced by the **Communicator manager** about the current state of the network and the infrastructure and the specific condition that should be maintained, this component tries to generate the best component allocation. In this vision, due to the high complexity for the management of dynamic conditions, AI can be injected to support the decision of the best deployment strategy to adopt at run time.

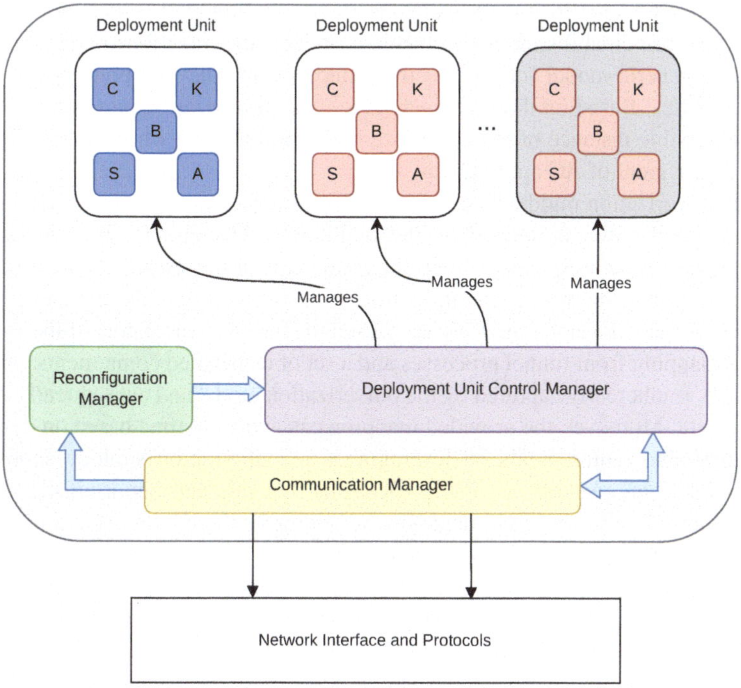

Fig. 3.5 Overall architecture of the pulverization middleware. The colored boxes on the bottom of the figure represent the main components of the middleware, and the light blue lines denote the relationship between components. Different colors of the components belonging to each "deployment unit" denote a different instance of a logical device. The different background colors of each "deployment unit" denote that device belongs to a different application. The box named *Network Interface and Protocols* represents the platform-specific interface and protocols made available by the host on which the middleware is executed

With this architecture, once the application is "pulverized," the middleware takes into account the execution of the component via the **Deployment unit control manager**, the management of the underlying infrastructure and the communications are demanded to the **Communication manager**, and finally, the **Reconfiguration manager** adapts the system to maintain specific QoS of the system providing real-time reconfiguration plan.

In Fig. 3.5, the described architecture and the relationship between the middleware components are depicted.

3.5.4 Fluidware and Pulverization Middleware

This section motivates how the pulverization middleware can be an instance of the Fluidware middleware model for CAS and CPS scenarios.

Section 3.4 exemplifies the principal traits for the Fluidware middleware, showing the five principal distinctive elements of it. The aforementioned characteristics refer to a generic model for an effective Fluidware middleware pointing common characteristics. Based on these characteristics, the pulverization middleware represents a possible instance of the Fluidware middleware model where the main focus is the deployment of AC applications.

The pulverization middleware satisfies the five objectives as follows. The funnel process specification is defined by the middleware leveraging a message-stream communication pattern, which allows the connection of the stream of data produced by the devices to the rest of the infrastructure demanded to elaborate and process the stream; then, the elaborated stream is used to control the actuators of the system.

The mapping from funnel processes and a set of distributed components, pointed out in (2), is inherently captured by the pulverization model and consequently by the middleware. Moreover, the provided mapping can evolve in time based on dynamic conditions and requirements by leveraging a reconfiguration strategy supported, i.e., by AI techniques. As a consequence, this dynamic and opportunistic system reconfiguration, allowing fluid deployment, addresses the points (3) and (4) of the main objectives of the Fluidware middleware model.

Since each middleware instance can manage multiple components belonging to multiple device in the system, privacy and security concerns can be addressed by defining appropriate *reconfiguration rules* to enforce the appropriate placement of component execution. In this way, also the objective expressed by (5) is addressed by the pulverization middleware.

3.6 Conclusion

In conclusion, the Fluidware architecture is mainly featured by its fully distributed and super-peer approach, and the pulverization middleware represents an effective implementation for the deployment of AC applications. Indeed, the high flexibility of the pulverization model joined with the architecture of the middleware makes the funnels definition straightforward and dynamically manageable. More details about a methodology for specifying how Fluidware funnels can be decomposed through the pulverization approach a toolchain for supporting the formal description and final deployment are reported in Chap. 6.

References

1. Alam, T.: Cloud-based IoT applications and their roles in smart cities. Smart Cities **4**, 1196–1219 (2021)
2. Aloi, G., Fortino, G., Gravina, R., Pace, P., Savaglio, C.: Simulation-driven platform for Edge-based AAL systems. IEEE J. Sel. Areas Commun. **39**, 446–462 (2020)

3. Ansari, N., Sun, X. Mobile edge computing empowers internet of things. IEICE Trans. Commun. **101**, 604–619 (2018)
4. Barbuto, V., Savaglio, C., Chen, M., Fortino, G.: Disclosing edge intelligence: a systematic meta-survey. Big Data and Cognit. Comput. **7**, 44 (2023)
5. Casadei R., Pianini, D., Placuzzi, A., Viroli, M., Weyns, D.: Pulverization in cyber-physical systems: engineering the self-organizing logic separated from deployment. Future Internet **12**(11), 203 (2020)
6. Casadei, R., Viroli, M., Audrito, G., Pianini, D., Damiani, F.: Engineering collective intelligence at the edge with aggregate processes. Eng. Appl. Artif. Intell. **97**, 104081 (2021)
7. Casadei, R., Fortino, G., Pianini, D., Placuzzi, A., Savaglio, C., Viroli, M.: A methodology and simulation-based toolchain for estimating deployment performance of smart collective services at the edge. IEEE Internet Things J. **9**(20), 20136–20148 (2022)
8. Fortino, G., Guerrieri, A., Russo, W., Savaglio, C.: Middlewares for smart objects and smart environments: overview and comparison. In: Internet Of Things Based On Smart Objects: Technology, Middleware And Applications, pp. 1–27 (2014)
9. Guinard, D., Trifa, V., Karnouskos, S., Spiess, P., Savio, D.: Interacting with the SOA-based internet of things: discovery, query, selection, and on-demand provisioning of web services. IEEE Trans. Serv. Comput. **3**, 223–235 (2010)
10. Laroui, M., Nour, B., Moungla, H., Cherif, M., Afifi, H., Guizani, M.: Edge and fog computing for IoT: a survey on current research activities and future directions. Comput. Commun. **180**, 210–231 (2021)
11. Mamei, M., Zambonelli, F.: Field-based coordination for pervasive multiagent systems. Springer Science & Business Media, New York (2006)
12. Murturi, I., Dustdar, S.: DECENT: a decentralized configurator for controlling elasticity in dynamic edge networks. ACM Trans. Internet Technol. (TOIT) **22**, 1–21 (2022)
13. Palade, A., Cabrera, C., White, G., Razzaque, M., Clarke, S.: Middleware for Internet of Things: a quantitative evaluation in small scale. In: 2017 IEEE 18th International Symposium On A World Of Wireless, Mobile And Multimedia Networks (WoWMoM), pp. 1–6 (2017)
14. Savaglio, C., Fortino, G.: A simulation-driven methodology for IoT data mining based on edge computing. ACM Trans. Internet Technol. (TOIT) **21**, 1–22 (2021)
15. Sharma, R., Prakash, S., Roy, P.: Methodology, applications, and challenges of WSN-IoT. In: 2020 International Conference On Electrical And Electronics Engineering (ICE3), pp. 502–507 (2020)
16. Taheri, J., Dustdar, S., Zomaya, A., Deng, S.: Distributed computing continuum systems. In: *Edge Intelligence: From Theory To Practice*, pp. 1–30 (2023)
17. Tzavaras, A., Petrakis, E.: Web of things functionality in IoT: a service oriented perspective. In: 2021 12th International Conference On Information, Intelligence, Systems and Applications (IISA), pp. 1–8 (2021)
18. Villari, M., Fazio, M., Dustdar, S., Rana, O., Jha, D., Ranjan, R.: Osmosis: the osmotic computing platform for microelements in the cloud, edge, and internet of things. Computer **52**, 14–26 (2019)

Chapter 4
Towards an IoT-Oriented Software Engineering Methodology

Stefano Mariani and Franco Zambonelli

4.1 Introduction

The possibility of enriching physical objects and places with wirelessly accessible sensing, computing, and actuating capabilities enables the "Internet of Things" (IoT) vision [4, 22, 61]. This defines a scenario in which everything in our physical and social worlds can become the node of a *large-scale situated network*, supporting coordinated actions to sense and control the physical world itself and to facilitate our interactions with it [16].

The dramatic future impact of IoT in society, health, industry, and commerce is already widely recognized [31, 54, 58, 60]. However, despite the great deal of worldwide research in the area, the technologies to make IoT a systematic reality are far from being assessed. Early researchers in the IoT area have mostly focused on communication issues and enabling interoperability [4]. Later, a great deal of effort has been devoted to promoting sound approaches to facilitate the integration of resources and services towards the provisioning of software-defined distributed services for the IoT. For instance, as in the so-called Web of Things (WoT) vision [28, 52], by promoting the provisioning of resources in an IoT network in terms of Web Services, thus making it possible to develop distributed and coordinated IoT services by using standard Web technologies. More recently, the abstraction of "Digital Twins" [37] (DTs)—a software abstraction continuously reflecting the properties of IoT devices—has emerged as a promising approach to attack the complexity of building IoT systems.

WoT and DTs approaches—possibly exploited in synergy [46]—are both definitely promising and will most likely represent keystone technologies in the future

S. Mariani (✉) · F. Zambonelli
University of Modena and Reggio Emilia, Reggio Emilia, Italy
e-mail: stefano.mariani@unimore.it; franco.zambonelli@unimore.it

F. Zambonelli et al. (eds.), *Fluidware*, Internet of Things,
https://doi.org/10.1007/978-3-031-62146-8_4

of IoT. Indeed, a number of different programming approaches [7, 18, 32, 33]) are being proposed to support the design and development of IoT systems and applications. Also, several proposals exist for middleware systems supporting the adaptive execution of IoT applications based on WoT and DTs [44, 62], there included the possibility to dynamically adapt the deployment of software components by moving them from the cloud to the edge and vice versa [8], as in the Fluidware approach.

However, a *common unifying methodological approach* supporting their software design and development, grounded on a common set of abstractions, models, and methodologies, is still missing. This undermines the possibility of promoting a *systematic and disciplined approach* towards the development of complex IoT systems and thus limits the unfolding of the full potential of the IoT vision.

Against this background, this chapter attempts at framing some key general-purpose issues related to the engineering of complex IoT systems and applications, by synthesizing the common characteristics of existing proposals and application scenarios. Such common characteristics are then used to identify the *key software engineering abstractions* around which the process of designing and developing IoT systems and applications could revolve, and via which to organize a set of guidelines towards the definition of a general methodology for engineering IoT systems.

To this end, the chapter introduces a specific case study scenario, yet representative of a larger class of IoT systems and applications. In particular, we consider the case of a big hotel with a conference centre. We assume the hotel infrastructure (e.g. lighting, heating, etc.) and its facilities (guest rooms, conference rooms, and their associated appliances and objects) to be densely enriched with connected sensors and actuators. In such a scenario, as will be discussed in the following, different actors can contribute to setting up IoT services to support both the hotel management and the activities of its guests, but this requires the development of such services and of the overall IoT software infrastructure to be grounded on a sound methodological approach.

The remainder of this chapter is organized as follows. Section 4.2 introduces some background concepts about IoT and WoT. On this basis, Sect. 4.3 introduces the key concepts and abstractions central to the engineering of IoT systems. Section 4.4 sketches the guidelines for a methodology for the design and development of IoT systems. Section 4.5 discusses related software engineering approaches, and Sect. 4.6 concludes.

4.2 Background

Most current IoT approaches envision a future where millions of sensors, actuators, and services operate in an orchestrated way, with the active contribution of humans' sensing, actuating, and reasoning capabilities. This section provides an overview of the background concepts that characterize most current IoT visions and some general engineering issues.

4.2.1 Things

The IoT vision includes a vast number of physical objects, places, and people.

Physical objects can be made trackable and controllable wirelessly by integrating them with low-cost electronic devices. At the lower end of the spectrum, RFID tags or Bluetooth beacons can be attached to any kind of object to enable tracking their positions and possibly associate some digital information with them. More advanced RFID and beacons can integrate environmental or motion sensors (i.e. accelerometers) to detect present and past activities associated with such objects. Objects can also be made *actionable* by enabling remote control of their configuration/status via proper digitally controlled actuators, and possibly autonomous by delegating them to direct their own activities.

For example, in the hotel scenario, RFID tags can be attached to objects in rooms (such as remote controllers) to detect their presence and location. A roll-up board in the conference room can be integrated with some kind of Arduino-link controller to enable controlling its rolling–unrolling via a mobile phone. Window obscuring systems could autonomously regulate lighting conditions depending on the kind of activities detected in the conference room. From this perspective, autonomous robots or robotized objects [1] can be considered the highest end of the spectrum of smart "things".

For places, similar to objects, simple RFID tags or beacons can be attached to places in the hotel to detect the proximity of people to those places or sense specific environmental conditions (at least for sensor-enriched tags and beacons). Remotely actionable parts of the environment such as windows, doors, and walls can also be made. In a hotel scenario, rooms can be designed such that all entities affecting environmental conditions (e.g. thermostats, windows) can be remotely controlled. In a hotel conference centre, actionable walls can dynamically change the shape and dimensions of meeting rooms depending on needs [43].

Concerning persons, they can be perceived as first-class entities of the overall IoT vision [3]. Simply by having a mobile phone, they can be somewhat sensed in their activities and positions and can be asked to act in the environment or supply further sensing [26]. In the hotel scenario, people's positions and activities can be continuously detected to manage any possible emergency situation efficiently. People also have a different role in exploiting the overall infrastructure of digitally enriched objects and places to get the best from their living in that environment. This includes interacting with the physical environment and enriching their social experience in that environment, as described in Sect. 4.2.3.

4.2.2 Software Infrastructures

To make IoT systems usable and capable of serving purposes, there is a need for software infrastructures (that is, an "IoT middleware" [5, 44, 57]) capable both of supporting the "glueing" together of different things, and of providing some means for stakeholders and users to access the IoT system and take advantage of its functionalities.

Concerning the glue, this involves a variety of technical issues:

- *Interoperability*. To enable a variety of heterogeneous things to interact with each other, a set of shared telecommunication protocols and data representation schemes must be put in place [42], other than means to identify things [48]. The study of these issues dates to the very early stages of IoT research, a number of different proposals exist, and the way towards assessed standards is well paved.
- *Semantics*. Beyond mere interoperability, common semantics for concepts must be defined to enable cooperation and integration of things [6]. Also for this issue, a number of proposals grounded on standard Web technologies, and defining ontologies and schemas specifically suited for the physically and socially embedded nature of the IoT exists [42].
- *Discovery, Group Formation, and Coordination*. Most of the functions provided in the context of an IoT system derive from the orchestrated access and exploitation of a variety of things, possibly involving a variety of users and stakeholders. For instance, in the hotel scenario, the need to configure a conference room for slide presentations requires involving the beam projector and the lighting system, as well as the conference organizers and the speakers. This requires means to discover and establish relations between things, and between things and humans [3, 11], as well as *coordinating* their activities also accounting for their social relations [2].
- *Context awareness and self-adaptation*. The large number of connected things that will define future IoT scenarios, and their inherent ephemerality, unreliability, and mobility (e.g. things such as chairs or flipboards in the hotel conference centre can come and go, can get moved around, and can be placed sometimes in corners with no wireless connections) makes it impossible to anticipate which things will be available and for how long during their exploitation. This requires that the mechanisms for discovery, group formation, and coordination are capable of dynamically adapting to the general context in which they act [63] and that mechanisms are provided for such groups to dynamically self-adapt on-the-fly, or possibly even *self-organize* in a context-aware way [39].
- *Opportunistic deployment*. This is the ability of a system to deploy and redeploy its components on a specific tier of the executing infrastructure—at the Edge, in the Fog, or to the Cloud—in the *most convenient* way, based on any property relevant for the application or for the benefit of non-functional properties [14, 25] For instance, JAMScript and Node-red [55] allow the programming of applications that dynamically migrate across the Edge-to-Cloud spectrum. This is done, in both cases, by means of a particular orchestration language and a runtime for supporting its execution.

Concerning how to access to the functionalities and capabilities of individual things by users, as well as the orchestration of the functionalities of groups of things, the scene is currently dominated by the WoT [28, 52] and DT visions [37]. The idea is to expose services and functionalities of individual things in terms of REST services, each associated with a URI, and to be simply accessed via HTTP GET and POST calls. This makes also possible to rely on assessed Web technologies as far as the discovery of things and provisioning of coordinated group services are concerned. More recently, DTs started to be interpreted as complementary to other rich programming abstractions, such as agents, so as to synergistically exploit their strengths to better deal with the typical issues of IoT deployments and endow the "things" with autonomy [35].

In the past few years, a variety of proposals for middleware infrastructures to support the provisioning of IoT services and applications appeared [7, 20, 32, 53, 62]. Besides the specificities of the different proposals, most of them rely on: some basic infrastructure to support the WoT approach (i.e. to expose things in terms of simple services); some means of supporting, according to some specific coordination models, the discovery of things (and of their associated services), and the coordinated activities of groups of things (typically to be made available in the form of a Web site or a mobile app [36]); some solutions to make services and applications capable of self-adapting and self-organizing in a context-aware and unsupervised way.

4.2.3 Services and Applications

With the term "IoT System" we generally refer to the overall set of IoT devices and the associated middleware devoted to managing their networking and interactions. Conceptually above an IoT system, a specific software can be deployed to orchestrate the activities of the system so as to provide:

- A number of specific *services*, that is, means to enable stakeholders and users to access and exploit individual things, and direct/activate their sensing/actuating capabilities, but also coordinated services that access groups of things and coordinate their sensing/actuating capabilities. For instance, in a conference room of the hotel, other than services to access and control individual appliances, one can think of providing a coordinated service that, by accessing and directing the lightning system, the light sensors, and the windows obscuring system, can modify the overall situation of the room from "presentation mode" to "discussion mode" and vice versa.
- A number of more general-purpose applications, intended as more comprehensive software systems meant to both regulate the overall functioning of an IoT system (or of some of its parts), so as to ensure specific overall behaviour of the system, and provide a harmonized set of services to access the system and (possibly) its configuration. In the hotel scenario, one can think of applications to control the overall heating and lighting systems, and giving to hotel clerks access to services to change the configuration of the associated parameters.

Clearly, depending on the specific scenario, one can think of IoT systems in which services may exist only confined within the context of some general application, but also of scenarios in which there are services that can be deployed as stand-alone software.

Given a specific IoT system, powered by a specific software infrastructure, a large variety of systems and applications can be designed and deployed over it, depending on the specific purpose it is intended to serve, and the specific classes of users that will somewhat exploit its services and applications. To make sure that the design and development of IoT services and applications fulfil expectations in a dependable way, though, a *disciplined and rigorous* approach to software analysis, design, and development is necessary. This is definitely a general concern in software engineering, which is made even more relevant by the pervasive nature of IoT systems and by their primary role in supporting our everyday activities.

4.3 Key Software Engineering Concepts and Abstractions

Based on the above overview of IoT issues, in this section, we synthesize the central concepts and abstractions around which the development of IoT systems (spanning analysis, design, and implementation) should be centred. Figure 4.1 graphically frames such concepts in a conceptual stack.

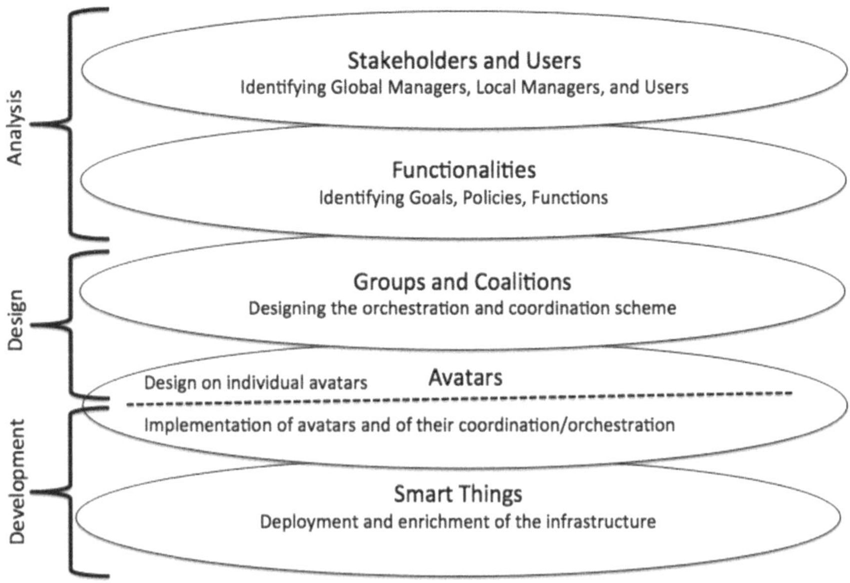

Fig. 4.1 The stack with the key concepts and abstractions for IoT systems engineering

4.3.1 Stakeholders and Users

The primary activities in the analysis of a system-to-be concern identifying the stakeholders and the users of the systems. That is, those persons/organizations who will own, manage, and/or use the system and its functionalities, and from which requirements should be elicited, aka the "actors".

In the case of IoT systems, there are three main classes of actors that can be identified:

- *Global Managers*: These are typically the owners of an overall IoT system and infrastructure, or at least persons empowered to exert control over the configuration, structure, and overall functioning of its applications and services. In the hotel scenario, the global manager corresponds to the hotel management, e.g. the system manager devoted to controlling the overall IoT system according to the directives of the hotel management. The global manager can establish the policies that the overall infrastructure and its applications must adhere to, e.g. for deciding heating levels or for surveillance strategies.
- *Local Managers*: These are typically owners (whether permanently or on a temporary basis) of a limited portion of the IoT system, or their delegates, and are empowered to enforce local control for that portion of the infrastructure. In the hotel scenario, these could correspond to hotel guests, which can be empowered to control the IoT system in their room, tune the local parameters, and exploit its services according to their own specific needs. Or they can be the organizers of a conference in charge of managing and configuring the services of the rented conference rooms.
- *Users*: These are typically persons or groups that have limited access to the overall configuration of the IoT applications and services, i.e. cannot impose policies but are nevertheless entitled to exploit its services. In the hotel scenario, these can include conference delegates that are authorized to access the conference facilities (e.g. uploading presentations in the projector and regulating the microphone and lighting in the meeting room) but are not entitled to modify the basic configuration of the infrastructure.

The three identified classes of actors are of a very general nature, regardless of the hotel scenario. For example, considering a scenario of energy management in a smart city, they could correspond to, respectively: city managers, house/shop owners, private citizens, and tourists. In the area of urban mobility, they could correspond to, instead: mobility managers, parking owners or car-sharing companies, and private drivers.

4.3.2 Functionalities

The development of IoT applications and services cannot simply reduce to understand the functionalities that objects or groups of objects have to provide but has to account for a more comprehensive approach, considering the following facts:

- Besides things provided with basic sensing/actuating functionalities, one should consider the presence of smarter things that can be activated to perform autonomously some long-term activities associated with their nature and role in the socio/physical environment in which they situate. These can range from simple cleaning robots to more sophisticated autonomous personal assistants [1].
- IoT applications are not simply concerned with providing a suite of coordinated functionalities, but they should also globally regulate the activities of the IoT systems on a continuous basis, according to the policies established by its stakeholders and to their objectives.

As a consequence, developing IoT services and applications, other than defining and implementing functionalities, most often implies defining *policies* and *goals* associated with services and applications [59]. In general terms, policies and goals represent a desirable "state of affairs" to strive for. In the context of an IoT system, policies and goals represent specific configurations of the overall socio/cyber/physical system (or of a portion of it) that IoT applications and services are in charge of eventually producing and/or maintaining, respectively.

In this perspective, the overall functionalities to be defined by IoT applications and services can be framed as follows:

- *Policies*. These express *desirable permanent configurations or states* of functioning of an overall IoT system (global policies) or portions of it (local policies) and have the aim of regulating the overall underlying IoT system. In the hotel scenarios, global policies can be defined, e.g. to specify the maximum occupancy levels in each room, and have this monitored by local cameras in order to invite people to move into different rooms whenever needed. Policies are meant to be *always active and actively enforced*. Although, from the software engineering viewpoint, the focus is mostly on application-level policies, policies can also account for the proper configuration of the underlying hardware and network infrastructures, in line with a software-defined networking perspective [30]. The definition of global and local policies is generally in charge of the global managers, although local managers can also be entitled to set up temporary local policies on local portions of the system (provided they do not contrast with the ones imposed by the global managers).
- *Goals*. These express desirable situations or states of affairs that, in specific cases, *can/should be activated*. The activation of a goal may rely on specific preconditions (i.e. the occurrence of specific events or the recognition of some specific configurations in the IoT system) or may also be triggered by user action (e.g. if the activation of a goal can be invoked as a service). The typical postcondition (deactivating the pursuit of a goal) is the achievement of the goal itself, although one can also consider postconditions for terminating the goal earlier than its achievement. In the hotel scenario, a natural example could be of activating an evacuation procedure upon detection of fire by some sensors (precondition), whose goal (and postcondition) is to achieve quick evacuation of all the people inside the building. To this end, the activation of a goal can trigger the activities of digital signages and controllable doors in order to rationally guide

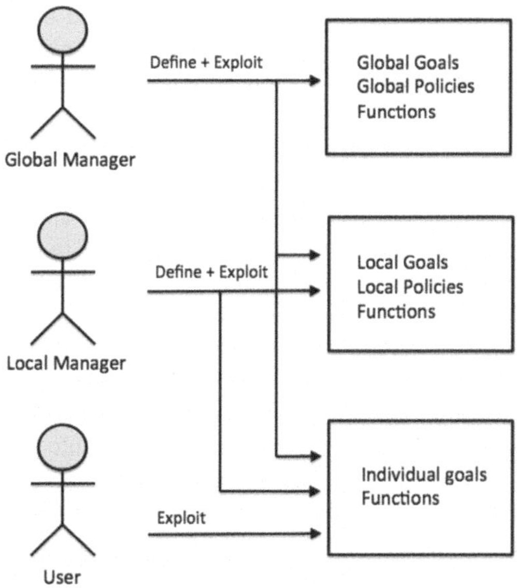

Fig. 4.2 The actors and the functionalities of IoT systems

people towards the exits. As for policies, the definition of global and local goals is generally in charge of global, and sometimes of local, managers.

- *Functions*. They define the sensing/computing/actuating capabilities of individual things or of a group of things, or the specific resources that are to be made available to managers and users in the context of specific IoT applications and services. Functions are typically made accessible in the form of services and can sometimes involve the orchestrated access to the functions of a multitude of individual things. In the hotel scenario, one can think of the individual functionalities of the appliances in a conference room (e.g. open/close a curtain, display slide/change slide in a projector), as well as more complex functionalities that can be achieved by orchestrating things (e.g. set up a room for a presentation by closing all curtains and switching off all lights). Functions and the associated services are typically defined by global and possibly local managers but are exploited also by everyday users of the IoT systems (e.g. the hotel guests and the conference attendees).

Figure 4.2 shows the different roles of IoT actors in defining and exploiting the functionalities of IoT systems.

4.3.3 Avatars and Coalitions

The "things" in an IoT system can correspond to a variety of different objects and ICT devices, other than to places and humans, each relying on a plethora of different technologies and capabilities. Accordingly, from both the viewpoints of the glueing software infrastructure and the software engineering practice, it is necessary to define higher-level abstractions to practically and conceptually handle the development of applications and services and to harmonically exploit all the components of the IoT system.

Most of the proposals for programming models and middleware acknowledge this need, by means of a software layer in which individual things are virtualized in some sort of software abstraction [28]. The WoT perspective, for instance, suggests abstracting things and their functionalities in terms of generic resources, to be accessed via RESTful calls [23, 24], possibly associating specific external HTTP gateways to individual things, whenever they cannot directly support HTTP interfacing [10]. Other approaches suggest adopting a more standard SOA or object-oriented approach [47]. Also, some proposals consider associating autonomous software agents to individual things [56] and resources [41], which we think well suits the fact that goals to be pursued in autonomy may be associated with things.

In addition, as already stated, some things make no sense as individual entities as far as the provisioning of specific services and applications is concerned and are to be considered part of a group thus delivering their services as a coordinated group. This applies both to the cases in which a multitude of equivalent devices must be collectively exploited, in order to abstract from the presence of the individuals and rather rely on the capabilities of the group [7, 9], and to the cases in which the functionalities of the group complement each other and need to be orchestrated [56, 57].

With these considerations in mind, and in an effort to synthesize from a variety of different proposals, we suggest the following unifying abstractions (See Fig. 4.3):

- *Avatars*. Borrowing the term from [38], we define an avatar as the general abstraction both for individual things and for a group of things that logically define a *unique concept*. Avatars abstract away from the specific physical/social/technological characteristics of the things they represent, as well as (for the case of groups) from the specific coordination techniques that are used to coordinate individual things within the group. Furthermore, depending on the specific responsibilities that an avatar would bear within the IoT system at hand, and on the programming paradigm and model adopted, the avatar may be realized as an agent, a DT, or even a Fluidware funnel process—e.g. an agent if cognitive abilities are needed for decision-making, a DT if what does matter the most is the coupling with the physical entity, and a funnel process if the focus must be on adaptive and deployment-agnostic data processing. Avatars are defined via:
  - *Identity*. An avatar has a unique identity and is addressable. An avatar enclosing a group of avatars does not necessarily hide the identities of inner avatars, but it has its own identity.
  - *Services*. Services represent access points for exploiting the peculiar capabilities of avatars. That is, depending on the kinds of things and functionalities it abstracts: triggering and directing the sensing/computing/actuating capabilities, or accessing some managed resources.

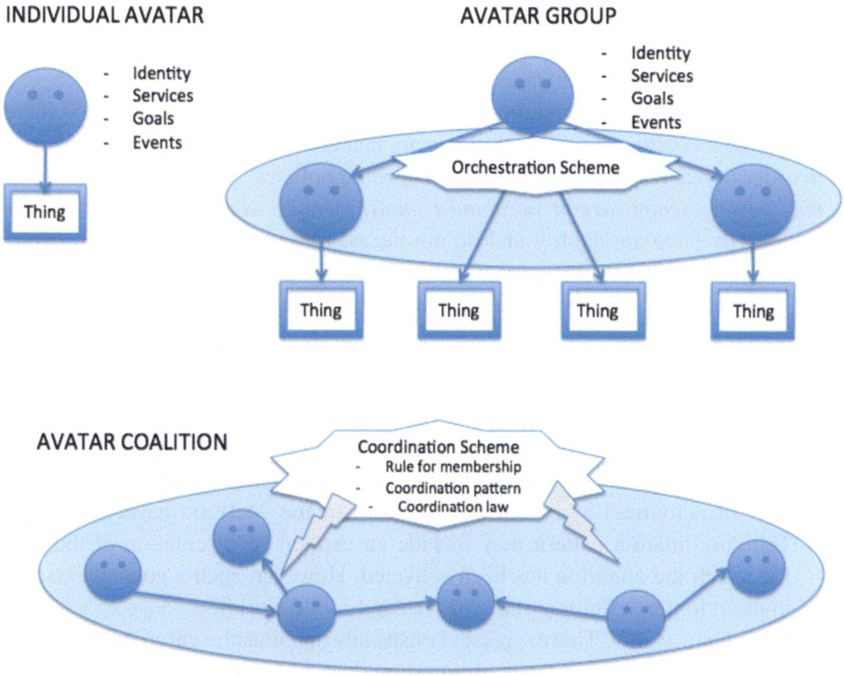

Fig. 4.3 Avatars, groups, and coalitions

- *Goals*. Goals, in the sense of the desired state of affairs, can be associated with Avatars. Goals may have a precondition for autonomous activation or may be explicitly activated by a user or by another avatar.
- *Events*. Events represent a specific state of affairs that can be detected by an avatar and that may be of interest to other avatars or to users. Other avatars or users can subscribe to events of interest.

Clearly, for groups of avatars, an internal *Coordination Pattern* must be defined to orchestrate the activities/functionalities of the objects (or of the other avatars) it includes. In general terms, coordination patterns define the internal workflow of activities among the composing objects and avatars, and the constraints/conditions they are subject to. Coordination patterns may also account for contextual information, thus making the activities of the group of avatars adaptive to changes in the context. The idea of defining coordination rules and constraints to coordinate the access and usage of things is generally adopted in most middleware and programming approaches for IoT [30, 47, 49, 62].

The above abstraction is perfectly in line with, and accounts for all the required characteristics of, the systems we already cited. The idea is not fully in line with that of RESTful Web-based approaches, however, because of the stateful concepts of goals and events. Nevertheless, the literature shows that most WoT approaches recognize the need to somehow incorporate similar concepts even within RESTful architecture [24], because they are necessary to suit the dynamic and contextual nature of IoT systems and applications.

- *Coalitions*. Borrowing the term from the area of multiagent systems [13], here we define a coalition as a group of avatars that coordinate each other's activities in order to reach specific goals or enact specific policies. Accordingly, *coalitions may be of a temporary or permanent nature*. Unlike avatars, coalitions do not necessarily have an identity and do not necessarily provide services. To define and bring a coalition into action, the abstraction of the coalition must be defined (at least) in terms of a *coordination scheme* that should include:
  - *Rules for membership*. These specify the conditions upon which an avatar should/could enter a coalition. From the viewpoint of individual avatars, the act of entering a coalition can be represented by the activation of a specific goal based on preconditions that correspond to the rules for membership.
  - *Coordination pattern*. This defines the pattern (interaction scheme and workflow of activities) by which the members of the coalition have to interact. The coordination pattern may include an explicit representation of the goal for which the coalition has been activated. However, such a goal can also be implicit in the definition of the pattern and of the workflow.
  - *Coordination law*. This expresses constraints that must be enforced in the way the avatars involved in the coalition should act and interact.

In addition, one can consider the possibility to subscribe to events occurring within the coalition.

The view of avatar coalitions can be of use to realize policies or to aggregate groups of avatars based on similarity, so as to make them work collectively without forcing them to specific orchestration. This is coherent with the idea of *aggregate programming* in sensor networks [9] and with spatial computing systems [7], to realize nature-inspired coordination schemes [19, 66], to enable the dynamic formation of groups focused on short-term goals [29, 50], or to define ensembles of services based on specific attributes [12, 17].

4.4 Towards a Software Engineering Methodology

As is the case with any new software paradigm and technological scenario, the successful and widespread deployment of complex software applications and services for the IoT requires not only the identification of the proper software engineering abstraction—which we have attempted in the previous section—but also the identification of an appropriate software engineering *methodology*—i.e. a set of guidelines revolving around such abstractions and facilitating engineers in developing such systems in a robust, reliable, and repeatable fashion.

The definition of a complete software engineering methodology is a very complex issue and should rely on a large body of real-world experience. Also, it

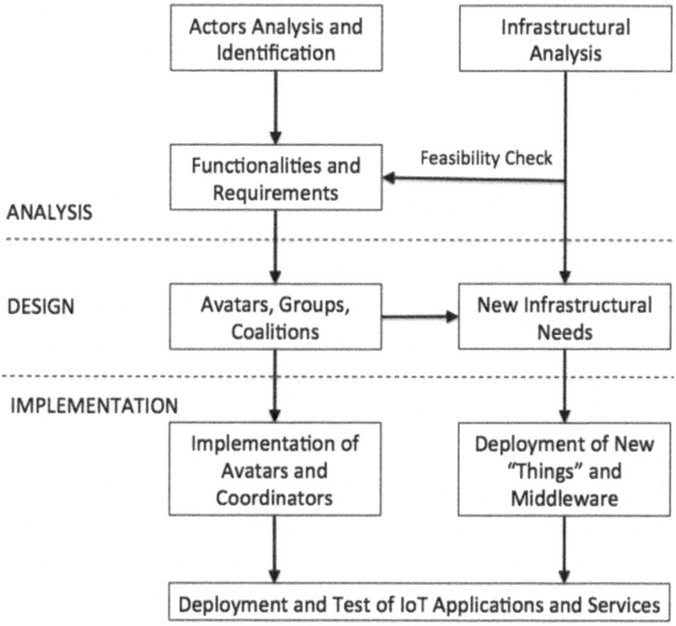

Fig. 4.4 Overview of a general IoT methodology

should be accompanied by a proper set of models and tools via which to represent and produce—in accordance with the basic abstractions of the paradigm—the conceptual and software artefacts that will eventually lead to the final product. Nevertheless, based on the current state of documented experiences, and given the analysis of the central abstractions provided above, it is possible to sketch some general guidelines and identify the different steps of the IoT software development process (see Fig. 4.4).

For the *analysis phase*, the envisioned activities include:

- *Actors analysis and identification*. Besides global managers, who are the primary stakeholders, this activity implies (in cooperation with the global managers) identifying the characteristics of the users of the systems, as well as of those users who can be entitled to play the role of local managers.
- *Infrastructure analysis*. Increasingly, in the future, new IoT applications and services will have to be deployed over an existing IoT hardware infrastructure and possibly on top of some associated middleware. Consequently, analysing the characteristics and limitations of such infrastructure will be a necessary prerequisite for the development of software over it.
- *Functionality and requirements*. These activities imply identifying goals, policies, and functions and define whether goals and policies are of a local or global nature. Requirements elicitation involves mostly global managers, who (besides having to define global goals and policies) are entitled to decide which services

to provide to end users and which local goals and policies can be set up by local managers. As in modern participatory approaches to system design [45], the involvement of potential end users and local managers could provide a more complete identification of needs and expectations. The identification of requirements should explicitly consider the characteristics of the infrastructure, in that some of the requirements may not be feasible on it, thus may require the later integration of some new component.

For the *design phase*, the envisioned activities include:

- *Design of Avatars, Groups, and Coalitions*. This must be defined with the goal of producing an overall design that is capable to realize the necessary goals, policies, and functions, in accordance with the requirements. We emphasize that the design of coalitions does not imply designing new avatars, but primarily designing the coordination scheme that will define and rule the activities of the coalition.
- *Identification of new infrastructural needs*. To understand what new devices or middleware services must be integrated into the infrastructure in order to realize the needed functionalities in accordance with the requirements.

For the *implementation phase*, the envisioned activities include:

- *Implementation of Avatars and Coordinators*. This clearly depends on the adopted middleware infrastructure and programming model. However, the abstraction of avatars and the concept of coordinators (intended as a software artefact to support a specific coordination scheme via which to realize coalitions) are of a very general nature and can apply to a large variety of actual IoT middleware and programming models.
- *Deployment of new things and new middleware characteristics*. This will take place according to the new infrastructural needs identified in the design phase. Of course, opportunistic deployment must be considered, but at the methodological level it is of low concern as it will be automatically handled by the middleware (which, instead, must be designed on purpose with that specific feature in mind).

Clearly, the methodology proposed above is not complete and requires the definition of models and tools to represent the different conceptual and software artefacts that are produced in the various steps (e.g. how we model an IoT requirement? Via which formal or semi-formal language can we represent the characteristics of an avatar or of a group? How can we turn such abstract representation into an implementation?). This is left for future work.

4.5 Related Work

In the past few years, research in the area of IoT exploded. Nevertheless, a few research works explicitly attacked the problem of defining new software engineering approaches specifically conceived for the IoT.

Some proposals for development frameworks for the IoT or for the WoT (whether middleware architecture [38, 44] or programming models [7, 21, 32]) are also accompanied by guidelines towards the development of applications. However, such guidelines are not grounded on general abstractions and have no general applicability besides the specific framework in which they are conceived. Similar considerations apply to the area of smart cities and urban computing [64], where middleware and programming approaches are being proposed—mostly of a special-purpose nature and focused on specific application scenarios such as participatory sensing [26, 27] or mobility management [34, 50]—but without accounting for the issue of defining general design and development methodologies.

Agent-Oriented Software Engineering (AOSE) research is strictly related to IoT engineering [65]. Indeed, AOSE tackles the problem of engineering large-scale systems and goal-oriented entities, possibly including robots [51] and humans [29] with conflicting goals and a multitude of stakeholders. This is somehow related to the IoT problems of accommodating services and a multitude of goals [56]. Indeed, the idea of goal-oriented groups we have introduced somehow borrows from the AOSE area. However, IoT requires the introduction of specific concepts and abstractions that AOSE, in general terms, does not address.

General approaches for the engineering of self-organizing computing systems have been proposed [19, 39, 40, 66, 67]. There, the key issue is to engineering complex distributed behaviours in large-scale systems lacking centralized control. These two characteristics are mostly shared by IoT systems, and indeed the problems of enabling the self-organization of specific behaviours have been outlined in the previous sections.

Mainstream software engineering researchers have recently put great attention to the problem of promoting self-adaptive features in software [15], to attack the problem of increased dynamic and unpredictability of operational environments. Such dynamics also affect IoT systems, in which the problem of ensuring continuity in functionalities requires the embedding of close control loops (along similar lines to those promoted in self-adaptive systems research) to continuously monitor the activities of the system and its environment, and eventually plan corrective actions.

4.6 Conclusions and Future Work

Despite the large number of research works that, at many levels, attempt at attacking specific problems related to the design and development of IoT applications and services, a general principled software engineering approach is still missing. This chapter, by having to frame the key conceptual abstractions revolving around the IoT universe, and by having sketched a methodology centred around these concepts, can represent the first small step towards a general discipline for engineering IoT systems and applications.

As IoT technologies mature, and real-world experiences accumulate, more research in the area of software engineering for IoT systems and applications will be needed, possibly exploiting contaminations with the relevant areas of AOSE [65] and software engineering for self-adaptive and self-organizing systems [15, 19].

References

1. Agrawal, H., Leigh, S.-W., Maes, P.: L'evolved: autonomous and ubiquitous utilities as smart agents. In: ACM International Joint Conference on Pervasive and Ubiquitous Computing, pp. 487–491. ACM, New York (2015)
2. Atzori, L., Carboni, D., Iera, A.: Smart things in the social loop: Paradigms, technologies, and potentials. Ad Hoc Netw. **18**, 121–132 (2014)
3. Atzori, L., Iera, A., Morabito, G.: From "smart objects" to "social objects": The next evolutionary step of the Internet of Things. IEEE Commun. Mag. **52**(1), 97–105 (2014)
4. Atzori, L., Iera, A., Morabito, G.: The internet of things: A survey. Comput. Netw. **54**(15), 2787–2805 (2010)
5. Bandyopadhyay, S., Sengupta, M., Maiti, S., Dutta, S.: A survey of middleware for internet of things. In: Azcan, A., Zizka, J., Nagamalai, D. (eds.) Recent Trends in Wireless and Mobile Networks. Communications in Computer and Information Science, vol. 162, pp. 288–296. Springer, Berlin (2011)
6. Barnaghi, P., Wang, W., Henson, C., Taylor, K.: Semantics for the internet of things: early progress and back to the future. Int. J. Semant. Web Inf. Syst. **8**(1), 1–21 (2012)
7. Beal, J., Pianini, D., Viroli, M.: Aggregate programming for the internet of things. IEEE Comput. **48**(9), 22–30 (2015)
8. Bellavista, P., Giannelli, C., Mamei, M., Mendula, M., Picone, M.: Application-driven network-aware digital twin management in industrial edge environments. IEEE Trans. Ind. Informatics **17**(11), 7791–7801 (2021)
9. Bicocchi, N., Mamei, M., Zambonelli, F.: Self-organizing virtual macro sensors. ACM Trans. Auton. Adapt. Syst. **7**(1), 1–28 (2012)
10. Bovet, G., Hennebert, J.: Offering web-of-things connectivity to building networks. In: ACM Conference on Pervasive and Ubiquitous Computing—Adjunct Publication, pp. 1555–1564. ACM, New York (2013)
11. Bronzino, F., Raychaudhuri, D.: Abstractions and solutions to support smart-objects in the future internet. In: Proceedings of the 2Nd Workshop on Experiences in the Design and Implementation of Smart Objects, SmartObjects '16, pp. 41–46. ACM: New York (2016)
12. Bures, T., Plasil, F., Kit, M., Tuma, P., Hoch, N.: Software abstractions for component interaction in the Internet of Things. Computer **49**(12), 50–59 (2016)
13. Cao, Y., Yu, W., Ren, W., Chen, G.: An overview of recent progress in the study of distributed multi-agent coordination. IEEE Trans. Industr. Inform. **9**(1), 427–438 (2013)
14. Casadei, R., Fortino, G., Pianini, D., Russo, W., Savaglio, C., Viroli, M.: A development approach for collective opportunistic edge-of-things services. Inf. Sci. **498**, 154–169 (2019)
15. Cheng, B.H.C., et al.: Software engineering for self-adaptive systems: A research roadmap. In Software Engineering for Self-Adaptive Systems. Lecture Notes in Computer Science, vol. 5525, pp. 1–26. Springer, Berlin (2009)
16. Conti, M., Das, S., Bisdikian, C., Kumar, M., Ni, L., Passarella, A., Roussos, G., Troster, G., Tsudik, G., Zambonelli, F.: Looking ahead in pervasive computing: challenges and opportunities in the era of cyber-physical convergence. Pervasive Mob. Comput. **8**(1), 2–21 (2012)

17. De Nicola, R., Latella, D., Lafuente, A.L., Loreti, M., Margheri, A., Massink, M., Morichetta, A., Pugliese, R., Tiezzi, F. and Vandin, A.: The SCEL language: Design, implementation, verification. In: Software Engineering for Collective Autonomic Systems—The ASCENS Approach, pp. 3–71. Springer, Berlin (2015)
18. Dias, J.P., Restivo, A., Ferreira, H.S.: Designing and constructing internet-of-things systems: An overview of the ecosystem. Internet of Things **19**, 100529 (2022)
19. Fernandez-Marquez, J.L., Di Marzo Serugendo, G., Montagna, S., Viroli, M., Arcos, J.: Description and composition of bio-inspired design patterns: a complete overview. Nat. Comput. **12**(1), 43–67 (2013)
20. Fortino, G., Guerrieri, A., Russo, W.: Agent-oriented smart objects development. In: IEEE 16th International Conference on Computer Supported Cooperative Work in Design, CSCWD 2012, May 23–25, 2012, Wuhan, China, pp. 907–912. IEEE, New York (2012)
21. Fortino, G., Savaglio, C., Spezzano, G., Zhou, M.C.: Internet of things as system of systems: A review of methodologies, frameworks, platforms, and tools. IEEE Trans Syst Man Cybern Syst **51**(1), 223–236 (2020)
22. Gubbi, J., Buyya, R., Marusic, S., Palaniswami, M.: Internet of things (IoT): A vision, architectural elements, and future directions. Futur. Gener. Comput. Syst. **29**(7), 1645–1660 (2013)
23. Guinard, D., Trifa, V.: Towards the web of things: Web mashups for embedded devices. In: Workshop on Mashups, Enterprise Mashups and Lightweight Composition on the Web (MEM 2009), in proceedings of WWW (International World Wide Web Conferences), Madrid, Spain (2009)
24. Guinard, D., Trifa, V., Mattern, F., Wilde, E.: From the internet of things to the web of things: Resource-oriented architecture and best practices. In: Uckelmann, D., Harrison, M., Michahelles, F. (eds.) Architecting the Internet of Things, pp. 97–129. Springer, Berlin (2011)
25. Guo, B., Yu, Z., Zhou, X., Zhang, D.: Opportunistic IoT: Exploring the social side of the internet of things. In: Proceedings of the 2012 IEEE 16th International Conference on Computer Supported Cooperative Work in Design (CSCWD), pp. 925–929 (2012)
26. Hachem, S., Pathak, A., Issarny, V.: Service-oriented middleware for large-scale mobile participatory sensing. Pervasive Mob. Comput. **10**, 66–82 (2014)
27. Harnie, D., D'Hondt, T., Boix, E.G., De Meuter, W.: Programming urban-area applications. In: Ossowski, S., Lecc, P. (eds.) Proceedings of the ACM Symposium on Applied Computing, SAC 2012, pp. 1516–1521, Riva, Trento, Italy, March 26–30. ACM (2012) https://doi.org/10.1145/2245276.2232018. https://dblp.org/rec/conf/sac/HarnieDBM12.bib
28. Heuer, J., Hund, J., Pfaff, O.: Toward the web of things: Applying web technologies to the physical world. Computer **48**(5), 34–42 (2015)
29. Jennings, N.R., Moreau, L., Nicholson, D., Ramchurn, S., Roberts, S., Rodden, T., Rogers, A.: Human-agent collectives. Commun. ACM **57**(12), 80–88 (2014)
30. Kathiravelu, P., Sharifi, L., Veiga, L.: Cassowary: Middleware platform for context-aware smart buildings with software-defined sensor networks. In: Proceedings of the 2Nd Workshop on Middleware for Context-Aware Applications in the IoT, M4IoT 2015, pp. 1–6. ACM, New York (2015)
31. Kott, A., Swami, A., West, B.J.: The internet of battle things. Computer **49**(12), 70–75 (2016)
32. Latronico, E., Lee, E.A., Lohstroh, M., Shaver, C., Wasicek, A., Weber, M.: A vision of swarmlets. IEEE Internet Comput. **19**(2), 20–28 (2015)
33. Mainetti, L., Mighali, V., Patrono, L.: A software architecture enabling the web of things. IEEE Internet Things J. **2**(6), 445–454 (2015)
34. Mariani, S., Cabri, G., Zambonelli, F.: Coordination of autonomous vehicles: taxonomy and survey. ACM Comput. Surv. **54**(1), 19:1–19:33 (2022)
35. Mariani, S., Picone, M., Ricci, A.: Agents and digital twins for the engineering of cyber-physical systems: opportunities, and challenges. In: Annals of Mathematics and Artificial Intelligence (2023)
36. Mazhelis, O., Tyrvainen, P.: A framework for evaluating internet-of-things platforms: application provider viewpoint. In 2014 IEEE World Forum on Internet of Things (WF-IoT), pp. 147–152. IEEE, New York (2014)

37. Minerva, R., Lee, G.M., Crespi, N.: Digital twin in the IoT context: A survey on technical features, scenarios, and architectural models. Proc. IEEE **108**(10), 1785–1824 (2020)
38. Mrissa, M., Medini, L., Jamont, J.-P., Le Sommer, N., Laplace, J.: An avatar architecture for the web of things. IEEE Internet Comput. **19**(2), 30–38 (2015)
39. Omicini, A., Viroli, M.: Coordination models and languages: From parallel computing to self-organisation. Knowl. Eng. Rev. **26**(1), 53–59 (2011)
40. Parunak, V.: Go to the ant: engineering principles from natural multi-agent systems. Ann. Oper. Res. **75**, 69–101 (1997)
41. Pötter, H.B., Sztajnberg, A.: Adapting heterogeneous devices into an IoT context-aware infrastructure. In: Proceedings of the 11th International Symposium on Software Engineering for Adaptive and Self-Managing Systems (SEAMS '16), pp. 64–74. ACM, New York (2016)
42. Perera, C., Zaslavsky, A., Christen, P., Georgakopoulos, D.: Context aware computing for the internet of things: A survey. IEEE Commun. Surv. Tutorials **16**(1), 414–454 (2014)
43. Phillips, M.: The Slothbot Moving Wall Projects. http://arch-os.com/projects/slothbot/
44. Picone, M., Mamei, M., Zambonelli, F.: A flexible and modular architecture for edge digital twin: implementation and evaluation. ACM Trans. Internet Things **4**(1), 8:1–8:32 (2023)
45. Plattner, H., Meinel, C., Leifer, L.: Design Thinking: Understand–Improve–Apply. Springer Science & Business Media, Berlin (2010)
46. Ricci, A., Croatti, A., Mariani, S., Montagna, S., Picone, M.: Web of digital twins. ACM Trans. Internet Techn. **22**(4), 101:1–101:30 (2022)
47. Sarkar, C., Nambi, S.N.A.U., Prasad, R.V., Rahim, A.: A scalable distributed architecture towards unifying IoT applications. In: IEEE World Forum on Internet of Things, pp. 508–513 (2014)
48. Sarma, A.C., Girão, J.: Identities in the future internet of things. Wirel. Pers. Commun. **49**(3), 353–363 (2009)
49. Sarray, I., Ressouche, A., Gaffé, D., Tigli, J.-Y., Lavirotte, S.: Safe composition in middleware for the Internet of Things. In: Proceedings of the 2nd Workshop on Middleware for Context-Aware Applications in the IoT, pp. 7–12. ACM, New York (2015)
50. Sassi, A., Zambonelli, F.: Coordination infrastructures for future smart social mobility services. IEEE Intell. Syst. **29**(5), 78–82 (2014)
51. Schurr, N., Marecki, J., Tambe, M., Scerri, P.: Towards flexible coordination of human-agent teams. Multiagent Grid Syst. **1**(1), 3–16 (2005)
52. Sciullo, L., Gigli, L., Montori, F., Trotta, A., Di Felice, M.: A survey on the web of things. IEEE Access **10**, 47570–47596 (2022)
53. Singh, J., Thomas, F.J.-M., Pasquier, T., Bacon, J., Powles, J.E., Diaconu, R., Eyers, D.M.: Big ideas paper: policy-driven middleware for a legally-compliant internet of things. In: Proceedings of the 17th International Middleware Conference, Trento, Italy, December 12–16, 2016, p. 13 (2016)
54. Singh, R.P., Javaid, M., Haleem, A., Suman, R.: Internet of things (IoT) applications to fight against covid-19 pandemic. Diabetes Metab. Syndr. Clin. Res. Rev. **14**(4), 521–524 (2020)
55. Sosa, R., Kiraly, C., Rodriguez, J.D.P.: Offloading execution from edge to cloud: a dynamic node-red based approach. In: 2018 IEEE International Conference on Cloud Computing Technology and Science (CloudCom), pp. 149–152. IEEE, New York (2018)
56. Spanoudakis, N., Moraitis, P.: Engineering ambient intelligence systems using agent technology. IEEE Intell. Syst. **30**(3), 60–67 (2015)
57. Teixeira, T., Hachem, S., Issarny, V., Georgantas, N.: Service oriented middleware for the internet of things: a perspective. In: Proceedings of the 4th European Conference on Towards a Service-based Internet (ServiceWave'11), pp. 220–229. Springer, Berlin (2011)
58. Tun, S.Y.Y., Madanian, S., Mirza, F.: Internet of things (IoT) applications for elderly care: a reflective review. Aging Clin. Exp. Res. **33**, 855–867 (2021)
59. Van Lamsweerde, A.: Goal-oriented requirements engineering: A guided tour. In: Fifth IEEE International Symposium on Requirements Engineering, pp. 249–262. IEEE, New York (2001)
60. Varga, P., Peto, J., Franko, A., Balla, D., Haja, D., Janky, F., Soos, G., Ficzere, D., Maliosz, M., Toka, L.: 5g support for industrial IoT applications—challenges, solutions, and research gaps. Sensors **20**(3), 828 (2020)

61. Vermesan, O., Friess, P., Guillemin, P., Giaffreda, R., Grindvoll, H., Eisenhauer, M., Serrano, M., Moessner, K., Spirito, M., Blystad, L.-C., et al.: Internet of things beyond the hype: Research, innovation and deployment. In: Building the Hyperconnected Society-Internet of Things Research and Innovation Value Chains, Ecosystems and Markets, pp. 15–118. River Publishers, Denmark (2022)
62. Yao, L., Sheng, Q.Z., Dustdar, S.: Web-based management of the Internet of Things. IEEE Internet Comput. **19**(4), 60–67 (2015)
63. Ye, J., Dobson, S., McKeever, S.: Situation identification techniques in pervasive computing: A review. Pervasive Mob. Comput. **8**(1), 36–66 (2012)
64. Zambonelli, F.: Toward sociotechnical urban superorganisms. IEEE Comput. **45**(8), 76–78 (2012)
65. Zambonelli, F., Omicini, A.: Challenges and research directions in agent-oriented software engineering. Auton. Agent. Multi-Agent Syst. **9**(3), 253–283 (2004)
66. Zambonelli, F., Omicini, A., Anzengruber, B., Castelli, G., De Angelis, F.L., Di Marzo Serugendo, G., Dobson, S., Fernandez-Marquez, J.L., Ferscha, A., Mamei, M., Mariani, S., Molesini, A., Montagna, S., Nieminen, J., Pianini, D., Risoldi, M., Rosi, A., Stevenson, G., Viroli, M., Ye, J.: Developing pervasive multi-agent systems with nature-inspired coordination. Pervasive Mob. Comput. **17**(Part B), 236–252 (2015)
67. Zambonelli, F., Viroli, M.: A survey on nature-inspired metaphors for pervasive service ecosystems. J. Pervasive Comput. Commun. **7**, 186–204 (2011)

Chapter 5
A BPMN-Based Approach for IoT Systems Engineering

Flavio Corradini, Fabrizio Fornari, Sara Pettinari, Barbara Re, Lorenzo Rossi, and Francesco Tiezzi

5.1 Introduction

Nowadays, IoT systems enable more and more merging of physical and digital worlds by observing real phenomena based on sensing data and the consequent possibility of making decisions based on such data [41]. IoT systems are implemented in many domains, such as industry and home automation, healthcare, smart mobility, logistics, and so on [63] since they reduce the need for human intervention in activities that could be dangerous, repetitive, or infeasible.

The management of IoT systems is often challenging, requiring several advanced skills, especially when the goal is to ensure loosely coupled coordination of heterogeneous devices. These systems carry on computation activities that capture, manipulate, and route IoT data to coordinate the devices' activities toward achieving system (application or service) goals. This goes by the name of IoT process. IoT processes challenge the research community, asking for appropriate modeling notations, enactment approaches, and monitoring mechanisms [40].

In this regard, model-driven engineering (MDE) has been proven to be suitable for facilitating these activities since it permits abstracting from the heterogeneity of IoT platforms and IoT devices [12, 42, 43]. However, to make models machine-processable and, thus, accessible to automated analyses, syntheses, integration, and evolution, these models must conform to explicit modeling language [37], such as UML [65], SysML [34], etc. Along this direction, Business Process

F. Corradini · F. Fornari (✉) · S. Pettinari · B. Re · L. Rossi
School of Science and Technology, University of Camerino, Camerino, Italy
e-mail: Flavio.Corradini@unicam.it; Fabrizio.Fornari@unicam.it; Sara.Pettinari@unicam.it; Barbara.Re@unicam.it; Lorenzo.Rossi@unicam.it

F. Tiezzi
Dipartimento di Statistica, Informatica, Applicazioni, University of Florence, Florence, Italy
e-mail: francesco.tiezzi@unifi.it

F. Zambonelli et al. (eds.), *Fluidware*, Internet of Things,
https://doi.org/10.1007/978-3-031-62146-8_5

Management (BPM) turned out to be a well-established discipline that makes available modeling notations, enactment mechanisms, and monitoring techniques for the comprehensive management of business processes [27]. More in detail, concerning the modeling, BPM provides standard notations like the Business Process Model and Notation (BPMN) [58] to represent collaborations between IoT processes carried on by multiple devices [40, 52]. On the enactment, engines can directly interpret process models without requiring model-to-code translation. Finally, on the monitoring, Process Mining [71] is usually adopted to gather valuable insight from process executions [23].

This chapter is related to the ***device independence*** challenges addressed in the Fluidware project. Indeed, in Fluidware, whole collectives of IoT devices are abstracted as producers (or consumers, for actuators) of distributed streams of contextualized events, thus making system programming dependent on the availability of specific patterns of events, not on the specific identity of the devices producing or consuming them. More in detail, we explore and discuss how to model, enact, and monitor IoT systems exploiting the BPMN notation in a case study concerning a smart agriculture IoT system. We present a BPMN-based MDE where to: (i) understand and define the customer's project-specific requirements in a clear and binding way; (ii) model in an abstract way the devices participating in the IoT process, and define which activities those devices perform; (iii) enact the behavior of the IoT devices with a BPMN engine able to interpret BPMN models directly; and (iv) monitor the running system through insight gathered by process mining. Finally, based on the study, we point to possible directions for future work.

The rest of the chapter is organized as follows. Section 5.2 reports about related work. Section 5.3 is about the conducted study. Section 5.4 presents the requirement definition phase, Sect. 5.5 introduces the modeling and refinement phase, Sect. 5.6 discusses the enactment phase, and Sect. 5.7 presents the monitoring phase. Section 5.8 introduces open challenges, and Sect. 5.9 concludes the chapter.

5.2 Related Work

In this section, we discuss and motivate the adoption of MDE and of BPM techniques for the engineering of large-scale IoT systems.

5.2.1 Model-Driven Engineering for IoT

Adopting an MDE approach in developing IoT solutions is gaining attention since it enables the abstraction of technical details through models that can be successively used as a base for further refinement and to derive executable artifacts [35] possibly. The pivotal point of MDE consists of defining the application concepts and relationships through models, which can be analyzed and then used to develop software systems [67] abstracting from the complexity of an IoT application. Models are used to abstract from the heterogeneity of the technological solution, providing a common level where the main and most common application concepts can be represented. Refining those models then guides the user in the software's automatic or semiautomatic development. To support MDE approaches, several proposals rely on DSLs related to the IoT domain, providing a clear vision of their advantages and drawbacks [66]. Indeed, DSLs are provided to represent a specific IoT domain, as the Smart City scenario [68], or to design and simulate an entire IoT application targeting a specific software solution, as is the case of FIWARE [3]. Several works focus on the interoperability between heterogeneous IoT platforms used to execute IoT processes [60]. Some contributions propose model-driven approaches for developing cross-platform IoT applications and solving applications' portability across multiple heterogeneous IoT platforms proposing specific DSML for IoT [16, 17].

An MDE approach can provide support in modeling solutions for single-device development and for more complex and fully functional IoT solutions [12, 24, 55–57]. Some contributions focus on providing customization and reusability supporting mechanisms within a fully operational MDE approach [15].

Much research has been conducted in the IoT domain on developing IoT applications through MDE approaches [12]. Different models are being used and extended to represent different perspectives of IoT solutions. UML diagrams are among the most used models to represent the structure and behavior of IoT devices. The ThingML [55] and MARTE [32] approaches broadly use UML models with different tools and methodologies for developing distributed IoT reactive embedded systems. From a business point of view, to represent the entire management of an IoT-aware process, models are designed with graphical notations such as BPMN, which can be adapted to model different application scenarios [13, 14, 69].

5.2.2 Requirement Definition and Modeling

In recent years, more effective integration of business processes and IoT-related topics has been recognized as a *win–win* situation, making it possible for both IoT and business processes to overcome their natural limitations [48]. Organizations' business processes profited from the adoption and integration of IoT capabilities, leading to the notion of "IoT-Aware business process" [6, 10, 11, 31, 38, 53] or IoT-Enhanced Business Processes [18, 69, 70].

Concerning *requirement definition and modeling*, several approaches use BPMN 2.0 as is [14]. In contrast, others found it helpful to enrich the modeling language with additional constructs to grasp aspects and characteristics of the IoT world that influence business process modeling. Those that use BPMN as is [5, 25, 59, 64] usually adopt labeling practices to refer to IoT concepts, such as: providing labels that identify IoT devices as process participants (e.g., Sensor Device, Actuator Device, Irrigation System, Gateway), labeling activities in such a way that they refer

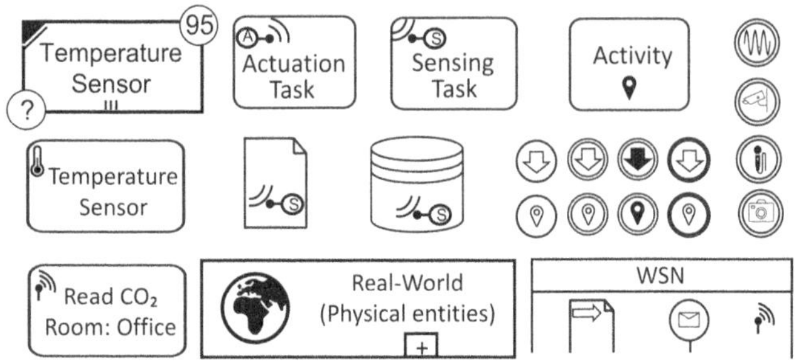

Fig. 5.1 BPMN IoT extended elements

to IoT performed activities (e.g., check temperature, activate ventilation, turn on the lights, etc.), labeling data objects with the type of data coming from IoT devices (e.g., GPS coordinates, Weights, Image), and using IoT data to label conditional flows (e.g., Temperature $>$ 30°C, Humidity $<$ 20%).

Several research works, instead, extend BPMN for integrating IoT concepts [44, 45, 53, 54, 75], see Fig. 5.1. The most common concepts that are introduced allow us to represent that IoT devices can participate in a business process and they can perform activities and generate events (i.e., sensing or actuating the environment), as well as produce data (also in the form of stream) that are used to guide decision-making. Some elements also introduce the possibility of representing mobile participants who can participate in a process and execute activities only in a certain location.

BPMN, whether extended or not, provides, on one hand, a notation that makes it easy to represent a process that abstracts from IoT devices and relies on IoT-generated data. On the other hand, BPMN also permits to define activities that the IoT device must carry on (in a certain sense to program the devices). Thus, BPMN enables a holistic approach that allows stakeholders to navigate through different levels of abstraction to use process models as efficient means to communicate about a process from different angles.

5.2.3 Enactment

Concerning the *enactment*, several different ways of executing BPMN processes integrating IoT concepts can be found. They can be categorized based on the location where the BPMN engine is deployed.

When the BPMN engine is deployed *on the Cloud* (or at a Fog level), the involved IoT devices only participate in the process by means of request/response or publish/subscribe mechanisms. When the BPMN engine is deployed *at the Edge* level, it can be used to execute processes that collect data provided by devices at a specific location (e.g., a room, a building, or a plaza), elaborate them to produce flows of commands, and possibly locally store or forward such data to the cloud.

Different ways exist to use BPMN to interact directly with IoT devices of different natures [9, 20]. When the BPMN engine is deployed *on IoT devices*, with enough computational power to support it, BPMN models can be directly executed on the IoT devices, defining the activities the device has to perform. Such activities can predigest and elaborate the data sensed by that device before forwarding it to other devices or collection points. They can also spread and redistribute as is collected data to nearby devices, realizing composite services with direct device-to-device coordination.

For low-level resource-constrained IoT devices that do not have enough computational power to support a BPMN engine, the BPMN model can be translated to the executable binary code, or a specific programming language, and executed on the IoT device. Generally, in this approach, the system relies on middleware technologies to translate the process model to the executable code and then send the code to the IoT devices for execution. As an example, in [26, 50] the authors model a scenario with standard BPMN and transform the model with recognition pattern techniques into executable code.

In the case of mere IoT sensors (resp., actuators) that can only sense and communicate data (resp., receive commands and actuate) and cannot elaborate such data (resp., be programmed), those can be configured to interact with an Edge device running a BPMN engine.

5.2.4 Monitoring

Considering *monitoring*, the implemented behavior of IoT systems is often a cumbersome task since a proper carry-out of an activity strictly depends on the capability to control several peripherals in real time and to manage its interactions with the environment [1]. In this regard, various methodologies can help to spot drawbacks in the IoT systems by monitoring their behavior, i.e., when and how activities have been done, to gather information useful for monitoring and optimizing the system. The most common approaches proposed in the literature range from formal verification [47], which is often unfeasible due to the system complexity, to quantitative analysis [1] such as field tests, simulations, or logging and playback [1], which requires humans in the constant observation system's execution. More recently, to overcome these drawbacks, some research works [4, 22] propose the use of process mining for monitoring the behavior of IoT devices through process models extracted from event logs generated during the execution of the systems.

Process mining is a well-established discipline for analyzing systems behavior that consists of many techniques for discovering, monitoring, and improving the actual processes. It was originally conceived as a solution for analyzing business process data, while more recently, the research community has started to apply it in other fields [8, 61]. Process mining relies on *event logs*, generated by the systems under investigation [71]. Event logs record data from systems' executions in an activity-centric manner, reporting at least two *attributes*, i.e., timestamp and activity name. Further attributes can be included in the events to provide context data that provide information about other perspectives. Events referring to the same system execution, i.e., *case*, are grouped into *traces* of events ordered by time. By analyzing such event logs, usually in the eXtensible Event Stream (XES)[1] format, process mining reconstructs the process model describing the system's behavior, spots bottlenecks and unwanted deviations, and gets insights on the execution performances such as duration and costs [71].

Despite its potential, process mining has not been extensively investigated due to the complexity of transforming IoT event data into event logs. Indeed, IoT devices produce continuous flows of data sent or received by their peripherals. Altogether, these event data shape the devices' behavior from the point of view of the peripherals usage in the form of sequences of acting and sensing events with a very low level of granularity and without correspondence with meaningful activities. Moreover, IoT devices enable the possibility of recording context data, e.g., spatial information, resource usage, and exchange of messages; therefore, they generate multiperspective event logs [29]. This relation between control flow and other perspectives leads to the need for novel process mining techniques that go beyond process discovery [73].

It is worth noticing that process mining is strongly focused on monitoring single-participant processes, while IoT systems collect many devices with distinct behaviors. More recently, several works [23, 28, 36, 51, 76] address the discovery of collaboration models to provide a comprehensive representation of the involved participants' behavior and interactions. The main challenges in applying process mining in this setting are related to the distributed nature of the recorded event data [23]. Since each participant executes part of the collaboration, events referring to the same collaboration execution are stored in different places as distributed event data. For a comprehensive analysis of the overall collaborative process, event data must provide the right information to analyze the entire traces of events characterizing a collaboration execution.

5.3 Toward a BPM Solution for IoT Process Management

This section provides an overview of the proposed approach, focusing on the involved phases, actors, and artifacts. The approach, Fig. 5.2, can be divided into four main phases: Requirements Definition, Model Design and Refinement, Model Enactment, and Monitoring. Our approach is BPMN-driven, as it adopts BPMN as

[1] https://xes-standard.org/

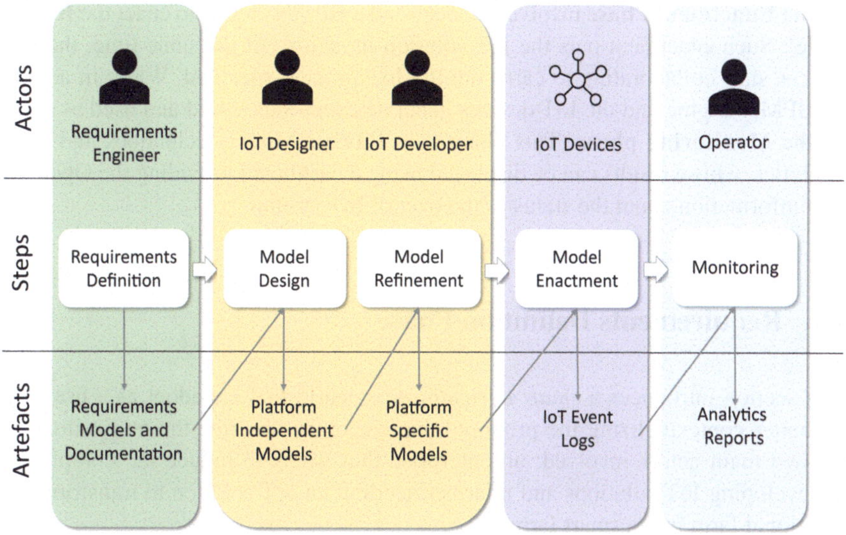

Fig. 5.2 BPMN-based model-driven engineering

a lingua franca for fostering communication between different actors participating in the IoT solution development process.

In the **Requirements Definition phase**, a requirements engineer identifies and understands the customer's project-specific demands and documents them in a clear and binding way. The requirements engineer can adopt graphical models to represent the functionalities the IoT solution has to support. In our approach, the requirements engineer adopts BPMN to design one or multiple drafts of the overall IoT solution requirements. Such models and textual documentation constitute the artifacts produced during the requirements definition phase and are to be considered input for the next phase of the approach.

The **Model Design and Refinement phase** involves the two roles of IoT Designer and IoT Developer. Based on the IoT solution requirements, the IoT Designer establishes, abstractly, the devices participating in the IoT process and defines which activities those devices perform. Such information is reported in a so-called Platform-Independent Model, designed with BPMN, that abstracts any technical information about the IoT devices used. Such details are handled by the IoT System Developer, who is aware of the specificities of the IoT devices to be used, what they can do, and how they communicate. In this phase, the IoT Designer and Developer interact closely; the same person can also play those roles. Once each Platform-Specific Model is defined, it is delivered to the right IoT device to enact such a model. In case a single model has been defined explicitly reporting all participants' activities, the model undergoes a splitting step that takes care of delivering the right activities that have to be performed to the suitable device.

The **Enactment Phase** involved devices and a BPMN engine to enact the BPMN model. Such enactment puts the IoT solution in action. At the same time, the IoT devices start collaborating to carry out the IoT process specified. While in action, the BPMN engine and the IoT devices generate event logs stored and used as input for the **Monitoring phase**. This phase uses Process Mining techniques and Data Analytics, whose results can be displayed using a dashboard providing the Operator with information about the status of the overall IoT system.

5.4 Requirements Definition Phase

This section introduces a *smart agriculture* scenario that we adopt as a use case for better contextualizing the proposed approach. We describe the motivations of the two main actors involved: an enterprise that wants to reduce the complexity of developing IoT solutions and a farmer needing an IoT solution to transform his traditional farm into a smart farm.

A farmer wants to profit from deploying IoT devices to automate some of the activities he conducts on his farm. He then contacts an enterprise adopting our approach. The client describes his needs to a requirements engineer. The client would like to adopt an IoT solution that could support him in automatizing the irrigation of cultures, assessing crop maturity levels, and automatically harvesting the ready cultures. The system should be composed of several static and movable IoT devices capable of collecting information and interacting with each other. The system's goal is to monitor data of interest for crop resiliency and sustainability, infer knowledge from these data, and make decisions for crop management. After manually inspecting the cultures, the farmer usually bases his irrigating and harvesting decisions on parameters such as weather conditions and soil status. The farmer would also like some help in predicting the maturity time of the cultures.

The requirements engineer schematizes the request done by the farmer in the BPMN model of Fig. 5.3. The BPMN model describes the need to collect different data that will be used to automatize the decisions to start/stop the irrigation or to harvest the mature cultures. Such data will also be stored for historical analysis and possible predictions. After the definition of the requirements, the client request will be taken into charge by the IoT department, which will start from the requirements that the requirements engineer defined and schematized in a BPMN model.

5.5 Model Design and Refinement Phase

The modeling phase of the IoT scenario is carried out in conjunction with an IoT Designer and an IoT Developer, who respectively take care of designing and developing the IoT solution.

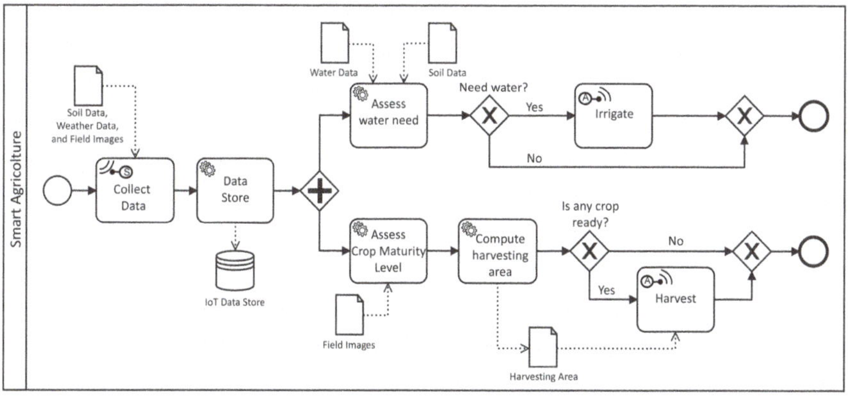

Fig. 5.3 BPMN model reporting the functionalities the IoT solution has to support

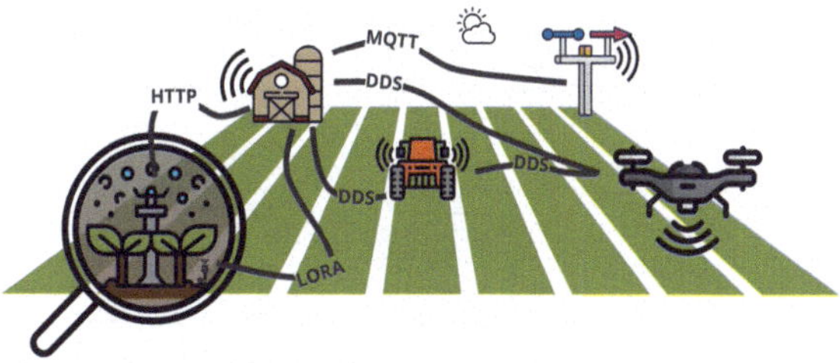

Fig. 5.4 IoT devices for the smart agriculture process

5.5.1 Designing the IoT Solution with BPMN

After analyzing the requirements, the IoT Designer identifies the IoT devices needed to develop the IoT system. This activity is necessary since it will contribute to the model definition by assigning activities to the devices, thus resulting in the model of the system's behavior and communication. Indeed, the devices able to implement the scenario could be different, e.g., the assessment of the crop maturity can be done either by a drone that takes pictures of the field from above or by a robot that scans the crop moving directly on the field.

We assume the IoT Designer identifies the following devices for the smart agriculture process: soil sensors, a smart irrigation system, a weather station, a drone, a smart tractor, and a resource manager. Figure 5.4 represents the system devices and the communication protocol they use to interact. **Soil sensors** fixed to the soil can collect and share information on the soil status, i.e., salinity and humidity levels. The **smart irrigation** fixed to the soil can be activated remotely. The **weather station** can gather information on current weather conditions. The **monitoring drone** can explore autonomously defined areas and collect images from above. The **harvesting tractor** can autonomously explore a working area and do the harvesting. The **resource manager** can communicate with the other devices, thus collecting information and triggering activities.

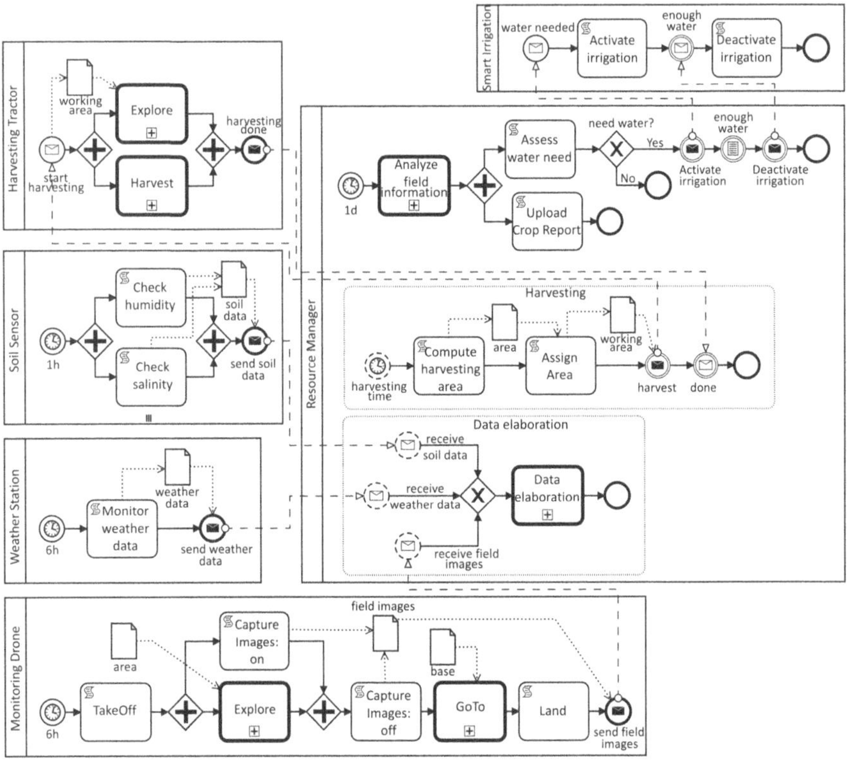

Fig. 5.5 BPMN collaboration of the smart agriculture IoT system

Now that the IoT devices have been identified, the IoT Designer can start to model the IoT system behavior, using the BPMN notation to design the activities and data exchange each IoT device will perform. Following the approach described in [5], the system behavior can be described through a unique collaboration diagram that encapsulates the processes of each IoT device. The designer can use the BPMN modeling tool she/he prefers as a supporting tool since all these tools rely on the standard *.bpmn* file format to ensure interoperability. One of the most popular and free tools is the Camunda modeler;[2] we used this tool for all the modeling activities shown next.

Figure 5.5 depicts a BPMN diagram representing one possible implementation of the smart agriculture process. The diagram contains six *pools* rectangles labeled with the name of the related device present in the system. They act as containers for the process to be carried out by the related device. The pool of the soil sensor is further decorated with a multi-instance marker (three vertical lines). It indicates that the system involves many devices that share the same behavior.

Each pool contains a *process* model, which describes the behavior of the devices. It consists of activities (rectangles with rounded corners), gateways (diamonds), and

[2] https://camunda.com/download/modeler/

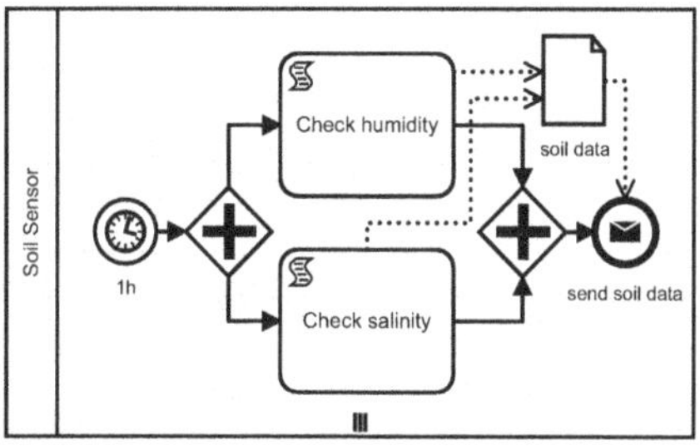

Fig. 5.6 Soil sensor process

events (circles) connected through sequence flow (solid arrows) and data objects (document icons) connected to activities and events through data associations (dashed arrows). For instance, the process of the soil sensor, Fig. 5.6, begins with a timer start event, which enacts a new process instance every hour. Then, the two activities nested in a block of parallel gateways indicated that the device measures humidity and salinity and collects these values concurrently. Finally, the retrieved soil information temporarily stored by the device is forwarded to the resource manager. This process is straightforward since it is associated with a device with few capabilities. A more intricate process is, for instance, related to the resource manager. Indeed, it collects data from other devices and acts as the de facto coordinator of the smart agriculture process. The resource manager can collect all the information produced by the devices to evaluate the need to irrigate the field and report the system status to the farmer. Moreover, it manages the harvesting phase. All these activities are reported in the corresponding BPMN process composed of a main process (the upper workflow in the pool) and two event subprocesses used for handling the management of the harvesting phase and reacting to the information produced by IoT devices. Specifically, the main process of the resource manager starts with a timer start event that enacts the execution once a day and performs a complex activity, expressed with a call activity, to evaluate the information retrieved from the field. Then, the resource manager performs two concurrent activities. The first one assesses the need to irrigate the field and, if so, it sends a message to activate the IoT smart irrigation and moves to a conditional event that evaluates when the irrigation should be stopped, and then, it sends another message to the smart irrigation to deactivate the irrigation. The second concurrent activity is responsible for collecting the field data and providing a report about the crop status. During the execution of the main resource manager process, two nonblocking event subprocesses can be triggered. A timer activates the harvesting subprocess start event, indicating the harvesting period. It defines the area that should be harvested and shares it with the tractor using a message event. After that, it moves to an idle state, waiting for a message that states the proper completion of the harvesting activity. Differently, the second subprocess can be enacted by three different message events. When the resource manager receives a message from a smart device, the content is processed to store the received information properly.

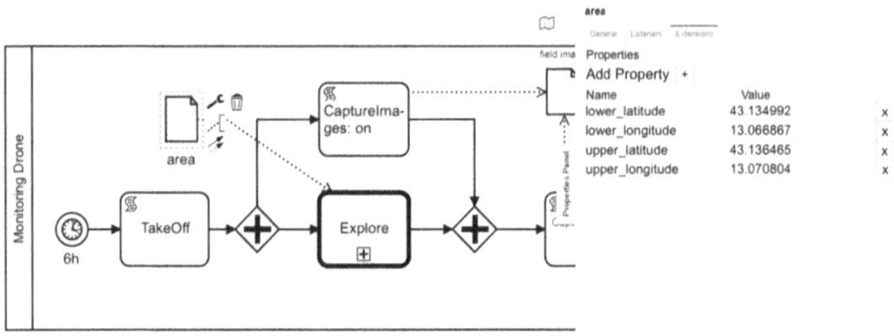

Fig. 5.7 Refinement of the *area* data object

5.5.2 Model Refinement

Once the IoT Designer has produced the BPMN model of the IoT system, it cooperates with the IoT Developer to properly configure the system and obtain an executable model. Indeed, the process model we got in the previous phase can be executed through dedicated engines that implement the execution semantics of BPMN. Moreover, the system participants with more computational capabilities, i.e., resource manager, monitoring drone, and harvesting tractor, can be integrated with a BPMN engine specifically customized to comply with the underlying architecture. There exist many implementations of BPMN engines. They range from commercial software targeting industrial scenarios, e.g., Camunda, Activiti, and jBPM, to lighter open-source flexible tools that devices with limited computational capabilities can run.

During the model refinement, the IoT Developer has to provide the BPMN file with additional details about communication, scripts, and variables:

- **Communication events**. BPMN message events have to be complemented by the IoT Developer with the information to be used at runtime to forward the message properly. First, a communication protocol must be chosen for sending and receiving events, e.g., DDS, Lora, or HTTP. Therefore, depending on the protocol, the IoT Developer must indicate a channel, an address, or a payload.
- **Script tasks**. Depending on the desired outcome, the IoT Developer has to include the lines of code to be executed at runtime by the corresponding device in each script task. For instance, considering the process of the monitoring drone in Fig. 5.5, the script task *CaptureImages:on* should contain the code that turns on the camera sensor and starts recording field images.
- **Data Objects**. Data objects can be containers for information used by the process elements. Therefore, the developer has to use such elements to instantiate the necessary variable names and values. Considering the process of the monitoring drone in Fig. 5.5, the data object *area*, depicted in Fig. 5.7, contains the declaration of the variables, describing the geographical boundaries of the area to be explored.

5.6 Enactment Phase

The enactment phase consists of the deployment and execution of the refined BPMN collaboration model in the IoT devices.

The first step in the deployment is splitting the BPMN collaboration into several process models, one for each participant, so that the devices can manage the execution independently. Depending on the computational capabilities of the device, the modeled behavior can be translated into code or interpreted by BPMN engines. In the former case, the BPMN model acts as a reference for the IoT Developer that has to directly implement the behavior using the languages imposed by the device manufacturer. For instance, the process that describes the weather station indicates how often the device checks the weather conditions and who is responsible for receiving the measured values. In the latter case, the device runs a BPMN engine. In such a way, the model acts as an actual source code; thus, no further model transformation is required to let the device execute the behavior. Every time an engine triggers a BPMN element that refers to an action to be done by the device (i.e., a task or a communication event), the code embedded in the element is evaluated at runtime by the underlying hardware. The other elements instead are entirely interpreted by the engine; for example, when an engine executes an XOR split gateway, it drives the control flow toward the sequence flow with the true condition without sending any command to the device.

At this point, once the IoT system starts, each device fires the execution of its part of the BPMN collaboration.

5.7 Monitoring Phase

The idea of using process mining to extract and analyze the behavior of IoT systems goes hand in hand with the BPMN-driven development described so far. Indeed, in our setting, the presence of BPMN engines that orchestrate the device activities enables the possibility of recording (and monitoring) related events. More in detail, the engines play the role of logging systems to produce event logs in the standard XES format.

During the system execution, every time an engine fires an activity, it records in the log an event that reports at least the timestamp, the activity name, and a case identifier. Other contextual data of interest can be included in the event depending on the available device information, e.g., an event corresponding to the drone's *TakeOff* activity could contain the GPS data, the battery level, the speed, and so on.

Since each device executes its behavior autonomously, one system execution results in one event log per device, each with specific attributes. Table 5.1 reports an excerpt of the event log of the smart agriculture collaboration. Specifically, Table 5.1a reports an excerpt of the log produced by the drone. In addition to standard attributes, like case identifier and activity name, the drone's position and

Table 5.1 Two event logs of the same system execution

(a) Drone's log excerpt				
Case	Activity	Timestamp	Position	Battery
001	TakeOff	09:33	{x:1, y:2, z:0}	100%
001	Explore	09:34	{x:1, y:2, z:5}	99%
001	CaptureImages: on	09:34	{x:1, y:2, z:5}	99%
001	CaptureImages: on	09:34	{x:12, y:-5, z:6}	38%
...	...	...	...	...
(b) Soil sensor's log excerpt				
Case	Activity	Timestamp	Soil data	Battery
010	Check humidity	09:01	{h:NaN, s:NaN}	88%
010	Check salinity	09:01	{h:0.5, s:NaN}	88%
010	Send soil data	09:02	{h:0.5, s:0.01}	88%
...	...	...	...	...

the battery level are stored. Table 5.1b refers instead to one of the soil sensors available in the field. In this case, the events report the sensed soil values and the battery level.

Altogether, the event logs produced by IoT devices open the possibility of applying several process mining techniques under certain assumptions.

Following the classical process mining paradigm [71], monitoring can be done vertically on single-event logs to observe their actual behavior (both online and postmortem), check compliance with system specifications, and enhance the monitoring dashboard with contextual data. In detail, process discovery[72] algorithms can extract from the logs the process model describing the behavior of the device under investigation. These algorithms exploit the causal relations between events to infer and generalize complex behavioral patterns such as parallelism, loops, choices, etc. Process discovery techniques produce, as a result, process models in different languages, i.e., BPMN, Petri Nets, WF-nets, and so on. The interested reader is referred to [2] for a detailed benchmark of these algorithms. To provide examples, we following present two models extracted from the log of a soil sensor using the process mining tools Disco[3] and PMTk.[4] These models give the Operator two complementary views of the behavior of the soil sensor. The first one, Fig. 5.8a, is a directly follows graph that directly represents the log. Indeed, it depicts the device executions using paths from the start to the end node. Specifically, it shows that, over 9982 cases, 4990 started with *Check humidity* and finished with *Check salinity*, while 4992 cases did the opposite. The BPMN model in Fig. 5.8b instead gives a generalized view of the process; using control-flow gateways, all the 9982 cases are grouped in the same parallel block.

[3] https://fluxicon.com/disco/

[4] https://pmtk.fit.fraunhofer.de/

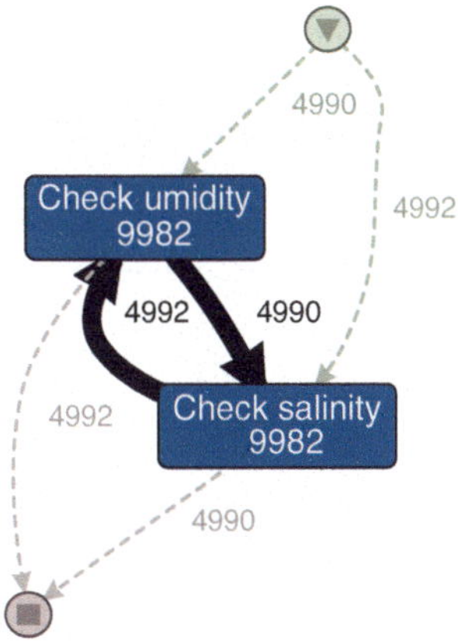

(a)

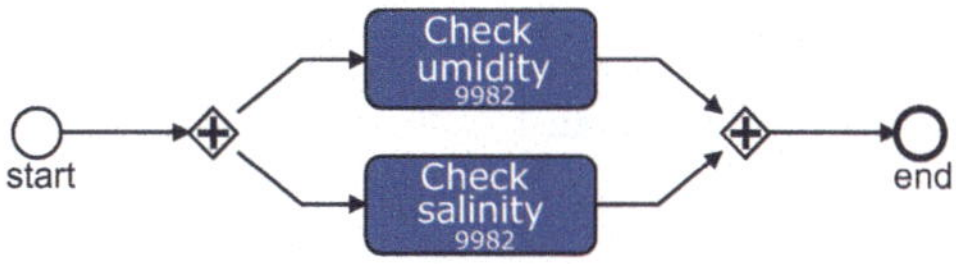

(b)

Fig. 5.8 Discovered behavioral models of the soil sensor. (**a**) Directly follows graph. (**b**) BPMN process model

Discovery is only the first step in enabling the monitoring of the IoT system. Indeed, the obtained models can be used to monitor ongoing process execution and spot deviations using token replay [71], i.e., the replay of execution traces on top of the process model. Furthermore, this technique can be coupled with live 3D replicas of the system [21]. Nevertheless, since the system involves multiple interacting devices, the Operator can benefit from process mining techniques tailored to monitor distributed systems. Indeed, interactions play a central role in distributed systems since they can influence the control flow of the single participants, i.e., a lost message may block a process waiting for it. For instance, Fig. 5.9 shows the BPMN collaboration diagram discovered using the Colliery tool described in [23]. Unlike the discovery of single-process models, the discovery of collaboration diagrams permits monitoring all system devices, including communication. Moreover, this technique provides analytics on message delivery to help the Operator to identify potential communication issues.

5.8 Discussion

Based on our collected results, we envision several directions for future research to bridge the gap between IoT and business processes. As discussed in the chapter, much is done referring to modeling IoT-aware business processes and how the modeling affects all the other phases of the business process life cycle [74]. However, there is no standard approach that can consider at the same time all the characteristics needed to support all the different phases. Indeed, for each phase,

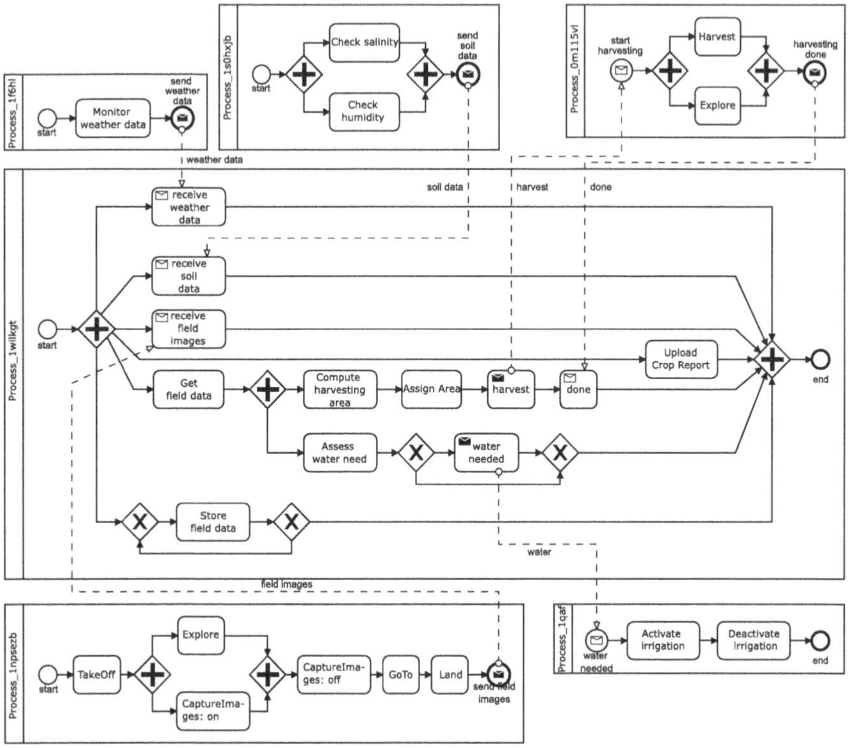

Fig. 5.9 BPMN collaboration of the smart agriculture system discovered using Colliery [23]

we see space for contributions. We discuss possible directions for future research, considering one phase at a time.

Process Modeling Referring to the modeling of business processes enhanced with IoT concepts, we extensively analyzed the literature through our investigation. As a result, we have shed light on the various proposals made over the years, discussing the lack of a standard way to model IoT-aware business processes. All the presented approaches, indeed, tackle the modeling of IoT-aware business processes from different perspectives, introducing specific graphical elements. The intervention of a well-recognized authority/group would be beneficial for defining a comprehensive and primarily used extension of BPMN for IoT. To start, graphical elements included in such a standard could be taken from the available literature, focusing on those attracting more attention or using the most (e.g., sensing and actuation tasks and smart data objects). However, a notation incorporating the best proposals from all approaches is still missing. Once standardized, such extension would help practitioners to design models that IoT-aware business process stakeholders can unambiguously understand.

Process Analysis Business processes interacting with the IoT world by automatizing the sensing and performing actions (using actuator devices) demand special attention. Such processes can affect the environment, and in many cases, it is hard (if not impossible) to reverse the actions they perform. Thus, there is a need for techniques to analyze, simulate, and test such processes before actually executing them. Already available methods and tools focusing on process analysis (e.g., [19, 39]), together with those referring to process simulation (e.g., [7, 33, 62]), could be adapted to handle the distinctive features of IoT-aware business processes.

Process Implementation Business process models are often used as input in process engines that support companies in organizing and optimizing their processes and the resources involved in such processes. To our knowledge, no engine presents specific functionalities for implementing IoT-aware business processes. Using already available process engines can be a starting point for process implementation. On one hand, this approach prevents anyone from "re-inventing the wheel" regarding business process implementation. On the other hand, it requires some compromises since process engines usually handle process models designed with precise notations (e.g., BPMN, Petri Net, BPEL). However, these notations have some expressiveness limitations. In particular, if a notation does not support IoT concepts such as sensors, actuators, and real-time and event streams, those concepts must be handled by the process designer with some workaround to make it still possible to use the process engine as it is. Our study showed no standard way to let already available process engines handle IoT-aware business processes. A survey of shared practices may be helpful. In addition, open-source process engines may be suitable for handling this limitation since they may allow for incorporating previously unforeseen aspects (e.g., using extensions). Moreover, much work has been done to provide users with IoT platforms that can ease the configuration and the usage of IoT systems [30]. Also, IoT platforms may benefit from an integration that aims to bridge the gap between tool support for IoT and tool support for business process management.

Process Monitoring IoT systems can be deployed to keep track of business process activities. Therefore, enhancing the support for monitoring IoT-aware business processes could allow business analysts to maintain the status of the monitored processes under control by leveraging the data retrieved by the IoT. For example, data can be reported and analyzed using dashboards, which are commonly used in IoT. The possibility of actually seeing the retrieved data and putting it into the business process activity that is being performed may empower business analysts with the support to spot erroneous or exceptional situations (e.g., an activity not being performed). This can enable process participants to perform near real-time corrective actions. In this respect, IoT data can permit to predict which activity will be completed next or the entire process result. In process mining, IoT data logs could be combined with activity logs to discover business processes based on historical data. This can enable retrospective analysis based on performance and detect process instances drifting from the specified behavior initially intended for the process. However, data provided by IoT systems and process engines differ in size and production rate. IoT data can quickly degenerate in Big Data, characterized by the well-known dimensions of volume, velocity, variety, and veracity [49]. Instead, data produced by process engines depend on the number of process activities to be performed, their kind (e.g., manual vs. automatized), and the speed required to achieve them. Handling these differences to make the best out of those data demands further research efforts. Mapping IoT data to process activities is a research area that is still under investigation [46].

5.9 Conclusion

This chapter presented several advantages the BPM domain can bring to managing IoT systems through concepts, techniques, and tools. In particular, we discussed three essential phases of the MDE and BPM: modeling, enactment, and monitoring with respect to IoT systems engineering. We observed the advantages of using a standard notation such as BPMN for guaranteeing interoperability between tools through the development phases. Besides modeling, BPMN is proven suitable for guiding the execution of the IoT process in a flexible and loosely coupled manner. Moreover, the BPMN notation impacts the monitoring of IoT processes since it can enhance standard IoT dashboards with the system's actual behavior via process discovery.

References

1. Afzal, A., Goues, C.L., Hilton, M., Timperley, C.S.: A study on challenges of testing robotic systems. In: Software Testing, Verification, and Validation, pp. 96–107. IEEE, New York (2020)
2. Augusto, A., Conforti, R., Dumas, M., Rosa, M.L., Maggi, F.M., Marrella, A., Mecella, M., Soo, A.: Automated discovery of process models from event logs: review and benchmark. Trans. Knowl. Data Eng. **31**(4), 686–705 (2018)
3. Barriga, J.A., Clemente, P.J., Hernández, J., Pérez-Toledano, M.A.: SimulateIoT-FIWARE: domain specific language to design, code generation and execute IoT simulation environments on FIWARE. IEEE Access **10**, 7800–7822 (2022)
4. Bertrand, Y., Weerdt, J.D., Serral, E.: A bridging model for process mining and IoT. In: Process Mining Workshops. LNBIP, vol. 433, pp. 98–110. Springer, Berlin (2021)
5. Bourr, K., Corradini, F., Pettinari, S., Re, B., Rossi, L., Tiezzi, F.: Disciplined use of BPMN for mission modeling of Multi-Robot Systems. In: Forum at Practice of Enterprise Modeling, vol. 3045, pp. 1–10. CEUR-WS.org (2021)
6. Brouns, N., Tata, S., Ludwig, H., Asensio, E.S., Grefen, P.: Modeling IoT-Aware business processes-a state of the art report. arXiv preprint arXiv:1811.00652 (2018)
7. Burattin, A., Re, B., Rossi, L., Tiezzi, F.: A purpose-guided log generation framework. In: Business Process Management. LNCS, vol. 13420, pp. 181–198. Springer, Berlin (2022)
8. Cairns, A.H., Gueni, B., Fhima, M., Cairns, A., David, S., Khelifa, N.: Process mining in the education domain. Int. J. Adv. Intell. Syst. **8**(1), 219–232 (2015)
9. Chang, C., Srirama, S.N., Buyya, R.: Mobile cloud business process management system for the internet of things: a survey. ACM Comput. Surv. **49**(4), 1–42 (2016)

10. Cheng, Y., Zhao, S., Cheng, B., Chen, J.: A service-based fog execution environment for the IoT-aware business process applications. In: Web Services, pp. 323–326. IEEE, New York (2018)
11. Cheng, Y., Zhao, S., Cheng, B., Chen, X., Chen, J.: Modeling and deploying IoT-aware business process applications in sensor networks. Sensors **19**(1), 111 (2019)
12. Ciccozzi, F., Spalazzese, R.: MDE4IoT: supporting the internet of things with model-driven engineering. In: Intelligent Distributed Computing X IDC 2016, vol. 678, pp. 67–76 (2016)
13. Compagnucci, I., Corradini, F., Fornari, F., Polini, A., Re, B., Tiezzi, F.: Modelling notations for IoT-aware business processes: a systematic literature review. Int. Workshops on Bus. Process. Manag. **397**, 108–121 (2020)
14. Compagnucci, I., Corradini, F., Fornari, F., Polini, A., Re, B., Tiezzi, F.: A systematic literature review on IoT-aware business process modeling views, requirements and notations. Softw. Syst. Model., **22**, 1–36 (2022)
15. Corradini, F., Fedeli, A., Fornari, F., Polini, A., Re, B.: FloWare: a model-driven approach fostering reuse and customisation in IoT applications modelling and development. In: Software and Systems Modeling **22**, (2022)
16. Corradini, F., Fedeli, A., Fornari, F., Polini, A., Re, B.: X-IoT: a model-driven approach for cross-platform IoT applications development. In: Symposium on Applied Computing, pp. 1448–1451. ACM, New York (2022)
17. Corradini, F., Fedeli, A., Fornari, F., Polini, A., Re, B., Ruschioni, L.: X-IoT: a model-driven approach to support IoT application portability across IoT platforms. Computing **105**(9), 1–25 (2023)
18. Fedeli, A., Fornari, F., Polini, A., Re, B., Torres, V., Valderas, P.: FloBP: a model-driven approach for developing and executing IoT-enhanced business processes. In: *Software And Systems Modeling*, pp. 1–30 (2024)
19. Corradini, F., Fornari, F., Polini, A., Re, B., Tiezzi, F., Vandin, A.: A formal approach for the analysis of BPMN collaboration models. J. Syst. Softw. **180**, 111007 (2021)
20. Corradini, F., Pettinari, S., Re, B., Rossi, L., Tiezzi, F.: A BPMN-driven framework for multi-robot system development. Robot. Auton. Syst. **160**, 104322 (2023)
21. Corradini, F., Pettinari, S., Re, B., Rossi, L., Tiezzi, F.: Executable digital process twins: towards the enhancement of process-driven systems. Big Data and Cognitive Computing **7**(3), 139 (2023)
22. Corradini, F., Pettinari, S., Re, B., Rossi, L., Tiezzi, F.: A methodology for the analysis of robotic systems via process mining. In: Enterprise Design, Operations, and Computing, pp. 117–133. Springer, Berlin (2023)
23. Corradini, F., Pettinari, S., Re, B., Rossi, L., Tiezzi, F.: A technique for discovering BPMN collaboration diagrams. In: Software and Systems Modeling (2024)
24. Costa, B., Pires, P.F., Delicato, F.C.: Modeling IoT applications with SysML4IoT. In: 42th Euromicro Conference on Software Engineering and Advanced Applications, SEAA, pp. 157–164. IEEE, New York (2016)
25. de la Croix, J.P., Lim, G.: Event-driven modeling and execution of robotic activities and contingencies in the Europa lander mission concept using BPMN. In: i-SAIRAS. ESA (2020)
26. Domingos, D., Martins, F.: Using BPMN to model Internet of Things behavior within business process. Inf. Syst. Proj. Manag. **5**(4), 39–51 (2017)
27. Dumas, M., La Rosa, M., Mendling, J., Reijers, H.: Fundamentals of Business Process Management. Springer, Berlin (2013)
28. Elkoumy, G., Fahrenkrog-Petersen, S., Dumas, M., Laud, P., Pankova, A., Weildich, M.: Secure multi-party computation for inter-organizational process mining. In: Business Process Modeling Development and Support. LNBIP, vol. 387, pp. 166–181. Springer, Berlin (2020)
29. Fahland, D.: Process mining over multiple behavioral dimensions with event knowledge graphs. In: Process Mining Handbook, pp. 274–319. Springer, Berlin (2022)
30. Fahmideh, M., Zowghi, D.: An exploration of IoT platform development. Inf. Syst. **87**, 1–47 (2020)

31. Fattouch, N., Lahmar, I.B., Boukadi, K.: IoT-aware business process: comprehensive survey, discussion and challenges. In: 29th IEEE Enabling Technologies: Infrastructure for Collaborative Enterprises, pp. 100–105. IEEE, New York (2020)
32. Faugere, M., Bourbeau, T., De Simone, R., Gerard, S.: MARTE: Also an UML profile for modeling AADL applications. In: 12th IEEE Engineering Complex Computer Systems, pp. 359–364 (2007)
33. Fracca, C., Bianconi, A., Meneghello, F., de Leoni, M., Asnicar, F., Turco, A.: BPSimpy: A python library for WfMC-standard process-simulation specifications. In: Dissertation, Doctoral Consortium, and Demonstration & Resources Track at Business Process Management, vol. 2973, pp. 97–101. CEUR-WS.org (2021)
34. Friedenthal, S., Moore, A., Steiner, R.: A Practical Guide to SysML: The Systems Modeling Language. Morgan Kaufmann, Burlington (2014)
35. Gascueña, J.M., Navarro, E., Fernández-Caballero, A.: Model-driven engineering techniques for the development of multi-agent systems. Eng. Appl. Artif. Intell. **25**(1), 159–173 (2012)
36. Hernandez-Resendiz, J.D., Tello-Leal, E., Marin-Castro, H.M., Ramirez-Alcocer, U.M., Mata-Torres, J.A.: Merging Event Logs for Inter-Organizational Process Mining, pp. 3–26. Springer, Berlin (2021)
37. Hölldobler, K., Rumpe, B., Wortmann, A.: Software language engineering in the large: towards composing and deriving languages. Comput. Lang. Syst. Struct. **54**, 386–405 (2018)
38. Hou, S., Zhao, S., Cheng, B., Cheng, Y., Chen, J.: Fragmentation and optimal deployment for IoT-Aware business process. In: Services Computing, pp. 657–664. IEEE, New York (2016)
39. Houhou, S., Baarir, S., Poizat, P., Quéinnec, P., Kahloul, L.: A first-order logic verification framework for communication-parametric and time-aware BPMN collaborations. Inf. Syst. **104**, 101765 (2022)
40. Janiesch, C., Koschmider, A., Mecella, M., Weber, B., Burattin, A., Di Ciccio, C., Fortino, G., Gal, A., Kannengiesser, U., Leotta, F., Mannhardt, F., Marrella, A., Mendling, J., Oberweis, A., Reichert, M., Rinderle-Ma, S., Serral, E., Song, W., Su, J., Torres, V., Weidlich, M., Weske, M., Zhang, L.: The internet of things meets business process management: A manifesto. IEEE Syst. Man Cybern. Mag. **6**(4), 34–44 (2020)
41. Khujamatov, K., Reypnazarov, E., Khasanov, D., Akhmedov, N.: Networking and computing in internet of things and cyber-physical systems. In: 14th Application of Information and Communication Technologies, pp. 1–6. IEEE, New York (2020)
42. Kirchhof, J.C., Kleiss, A., Rumpe, B., Schmalzing, D., Schneider, P., Wortmann, A.: Model-driven self-adaptive deployment of internet of things applications with automated modification proposals. ACM Trans. Internet Things **3**(4), 30:1–30:30 (2022)
43. Kirchhof, J.C., Rumpe, B., Schmalzing, D., Wortmann, A.: Montithings: model-driven development and deployment of reliable IoT applications. J. Syst. Softw. **183**, 111087 (2022)
44. Kirikkayis, Y., Gallik, F., Reichert, M.: Towards a comprehensive BPMN extension for modeling IoT-aware processes in business process models. In: 16th Research Challenges in Information Science. LNBIP, vol. 446, pp. 711–718. Springer, Berlin (2022)
45. Kozel, T.: BPMN mobilisation. In: World Scientific and Engineering Academy and Society, pp. 307–310. ACM, New York (2010)
46. Leotta, F., Marrella, A., Mecella, M.: IoT for BPMers. Challenges, case studies and successful applications. In: Business Process Management. LNCS, vol. 11675, pp. 16–22. Springer, Berlin (2019)
47. Luckcuck, M., et al.: Formal specification and verification of autonomous robotic systems: A survey. ACM Comput. Surv. **52**(5), 100:1–100:41 (2019)
48. Maamar, Z., Kajan, E., Guidara, I., Moctar-M'Baba, L., Sellami, M.: Bridging the gap between business processes and IoT. In: Database Engineering & Applications Symposium, pp. 2:1–2:10 (2020)
49. Marjani, M., Nasaruddin, F., Gani, A., Karim, A., Hashem, I.A.T., Siddiqa, A., Yaqoob, I.: Big IoT data analytics: Architecture, opportunities, and open research challenges. IEEE Access **5**, 5247–5261 (2017)
50. Martins, F., Domingos, D.: Modelling IoT behaviour within BPMN Business Processes. In: Procedia Computer Science, vol. 121, pp. 1014–1022. Elsevier, Amsterdam (2017)

51. Meroni, G., Baresi, L., Montali, M., Plebani, P.: Multi-party business process compliance monitoring through IoT-enabled artifacts. Inf. Syst. **73**, 61–78 (2018)
52. Meyer, S., Ruppen, A., Hilty, L.M.: The things of the internet of things in BPMN. In: Persson, A., Stirna, J. (eds.) Advanced Information Systems Engineering Workshops. LNBIP, vol. 215, pp. 285–297. Springer, Berlin (2015)
53. Meyer, S., Sperner, K., Magerkurth, C., Debortoli, S., Thoma, M.: Internet of Things architecture IoT-A project deliverable D2.2-concepts for modelling IoT-aware processes. Tech. rep., VDI/VDE Innovation and Technik GMBH, Germany (2012). eC FP7 IoT-A
54. Meyer, S., Sperner, K., Magerkurth, C., Pasquier, J.: Towards modeling real-world aware business processes. In: Web of Things, pp. 8–14. ACM, New York (2011)
55. Morin, B., Harrand, N., Fleurey, F.: Model-Based software engineering to tame the IoT jungle. IEEE Softw. **34**(1), 30–36 (2017)
56. Nepomuceno, T., Carneiro, T., Maia, P.H., Adnan, M., Nepomuceno, T., Martin, A.: *AutoIoT: A Framework Based on User-Driven MDE for Generating IoT Applications*, pp. 719–728 (2020)
57. Nguyen, X., Tran, H., Baraki, H., Geihs, K.: FRASAD: a framework for model-driven IoT application development. In: World Forum on Internet of Things, pp. 387–392 (2015)
58. OMG: Business Process Model and Notation (BPMN V 2.0) (2011)
59. Otsu, K., et al.: Supervised autonomy for communication-degraded subterranean exploration by a robot team. In: Aerospace Conference, pp. 1–9. IEEE, New York (2020)
60. Palau, C.E., Fortino, G., Montesinos, M., Exarchakos, G., Giménez, P., Markarian, G., Castay, V., Fuart, F., Pawłowski, W., Mortara, M., et al.: Interoperability of Heterogeneous IoT Platforms. Springer, Berlin (2021)
61. Pasquadibisceglie, V., Appice, A., Castellano, G., Malerba, D.: Predictive process mining meets computer vision. In: BPM, pp. 176–192. Springer, Berlin (2020)
62. Pufahl, L., Wong, T.Y., Weske, M.: Design of an extensible BPMN process simulator. In: Business Process Management Workshops. LNBIP, vol. 308, pp. 782–795. Springer, Berlin (2017)
63. Reggio, G., Leotta, M., Cerioli, M., Spalazzese, R., Alkhabbas, F.: What are IoT systems for real? an experts' survey on software engineering aspects. Internet of Things **12**, 100313 (2020)
64. Rey, R., Corzetto, M., Cobano, J.A., Merino, L., Caballero, F.: Human-robot co-working system for warehouse automation. In: ETFA, pp. 578–585. IEEE, New York (2019)
65. Rumpe, B.: Modeling with UML. Springer, Berlin (2016)
66. Salman, A.J., Al-Jawad, M., Al Tameemi, W.: Domain-specific languages for IoT: challenges and opportunities. In: Materials Science and Engineering Conference Series. Materials Science and Engineering Conference Series, vol. 1067, p. 012133 (2021)
67. Schmidt, D.C.: Model-driven engineering. Computer-IEEE Computer Society **39**(2), 25 (2006)
68. Sneps-Sneppe, M., Namiot, D.: On web-based domain-specific language for internet of things. In: 7th International Congress on Ultra Modern Telecommunications and Control Systems and Workshops, ICUMT, pp. 287–292 (2015)
69. Valderas, P., Torres, V., Serral, E.: Modelling and executing IoT-enhanced business processes through BPMN and microservices. J. Syst. Softw. **184**, 111139 (2022)
70. Valderas, P., Torres, V., Serral, E.: Towards an interdisciplinary development of IoT-enhanced business processes. Bus. Inf. Syst. Eng. **65**(1), 25–48 (2023)
71. van der Aalst, W.: Process Mining: Data Science in Action. Springer, Berlin (2016)
72. van der Aalst, W.: Foundations of Process Discovery, pp. 37–75. Springer, Berlin (2022)
73. Weerdt, J.D., Wynn, M.T.: Foundations of process event data. In: Process Mining Handbook, LNBIP, vol. 448, pp. 193–211. Springer, Berlin (2022)
74. Weske, M.: Business Process Management—Concepts, Languages, Architectures, 3rd edn. Springer, Berlin (2019)
75. Yousfi, A., De Freitas, A., Dey, A.K., Saidi, R.: The use of ubiquitous computing for business process improvement. Trans. Serv. Comput. **9**(4), 621–632 (2016)
76. Zeng, Q., Duan, H., Liu, C.: Top-down process mining from multi-source running logs based on refinement of petri nets. IEEE Access **8**, 61355–61369 (2020)

Chapter 6
Space-Fluid and Time-Fluid Programming

Danilo Pianini, Roberto Casadei, Stefano Mariani, Gianluca Aguzzi, Mirko Viroli, and Franco Zambonelli

6.1 Introduction

Fluidity is the property of fluids, namely substances that can flow easily and have the tendency of assuming the shape of their container. Viscosity is its antonym, which somewhat involves a resistance to change. Therefore, fluidity is a matter of *adaptiveness* in abstract space, and viscosity is where adaptation is difficult or hindered (e.g. by the need for manual intervention, lack of context awareness, or rigid architecture). What can it mean for computations to be fluid? In this chapter, we reply with a notion of computations that should move, with little resistance, where they are most needed and where they can be most effective and efficient for the application and constraints/preferences at hand. We thus cover two main kinds of computational fluidity that are investigated in the FluidWare project: *time fluidity* [26], namely the ability to adapt computation across time (e.g. to make it denser or sparser in specific time intervals, according to application needs); and *space fluidity* [10], namely the ability to adapt computation across space (e.g. by suitably distributing it across systems situated in environments where phenomena to track or react to are also spatially situated).

More specifically, regarding time, approaches to self-organisation [12, 24, 29] like field-based coordination [35] often assume that device computations are regulated by a periodic clock, configured on a per-application basis, and dictating a basic frequency of activity. In this chapter, we challenge this assumption and propose a

D. Pianini (✉) · R. Casadei · G. Aguzzi · M. Viroli
Università degli Studi di Bologna, Cesena, Italy
e-mail: danilo.pianini@unibo.it; roby.casadei@unibo.it; gianluca.aguzzi@unibo.it; mirko.viroli@unibo.it

S. Mariani · F. Zambonelli
Università di Modena e Reggio Emilia, Reggio Emilia, Italy
e-mail: stefano.mariani@unimore.it; franco.zambonelli@unimore.it

F. Zambonelli et al. (eds.), *Fluidware*, Internet of Things,
https://doi.org/10.1007/978-3-031-62146-8_6

reactive approach where scheduling follows the paths of change propagation. In particular, we enable *programming* of the scheduling in terms of *causality fields*, each providing a distributed notion of a computational "cause" (why and when a field computation has to be locally computed) and how it should change across time and space. Starting from low-level platform triggers, such causality fields can be organised into multiple layers, up to defining high-level, collectively computed time abstractions, to be used at the application level. This reinterpretation of the traditional view of time in terms of articulated causality relations allows us to express what we call "time-fluid" coordination, where scheduling can be finely tuned so as to select the triggers to react to, generally allowing to adaptively balance application performance (cf. system reactivity) and efficiency (cf. costs and resource usage).

Regarding space, we focus on a recurrent problem in pervasive computing systems, namely managing (i.e. estimating, predicting, or controlling) environmental signals that vary in space. Nodes can locally sense, process, and act upon signals and coordinate with neighbours to implement distributed collective strategies for the estimation of a spatial phenomenon through collaborative adaptive sampling of environmental data. Our proposal leverages the idea of dynamically partitioning space into competing regions that grow or shrink to provide accurate global-level sampling. Indeed, such regional partition defines a sort of virtualised space that is "fluid", as its shape adapts in response to pressure forces exerted by the underlying phenomenon. We provide an adaptive sampling algorithm in the field-based coordination framework, also proving it is self-stabilising and locally optimal, and showing it provides a means for fostering a tuneable trade-off between accuracy and efficiency.

The remainder of this chapter is organised as follows. Section 6.2 discusses previous research on adaptive computations across space and time; Sects. 6.3 and 6.4 present our field-based approach to time and space fluidity, respectively; and Sect. 6.5 provides concluding remarks.

6.2 State of the Art

Time fluidity relates computation adaptation to time, whereas space fluidity relates computation adaptation to space. In this section, we report on relevant previous research about time and space in distributed and pervasive computing systems.

6.2.1 Time and Synchronisation in Pervasive Computing Systems

Many algorithms exist for processing data in a distributed way, that is by a set of distributed processes [15]. However, when such data dynamically vary with time, processing must be periodically recomputed, which introduces the problem of identifying the appropriate frequency for computations. Most solutions typically adopt a predefined frequency [7, 15], but this strategy does not take into account that such frequency should be properly tuned to match the dynamics of the data being processed. Also, synchronising processes to let them organise processing in successive rounds is far from easy in asynchronous systems lacking a shared physical clock [11, 13]. Our proposal starts from similar problems, but is specifically conceived for distributed field-based data processing, and accounts for the strict relations between the spatial and temporal dimensions that exist in situated computations.

Specific timing problems also arise in the area of wireless sensor networks. There, acquiring a globally shared notion of time (as accurate as possible) is of fundamental importance [30] to capture accurate snapshots of the distributed phenomena under observation. However, global synchronisation also serves energy-saving purposes. In fact, when not monitoring or not communicating, the nodes of the network should go to sleep to avoid energy waste, but this implies that to exchange monitoring information with each other they must periodically wake up in a synchronised way. Several proposals exist for adaptive synchronisation in wireless sensor networks [1, 14, 16], dynamically changing the sampling frequency (and hence frequency of communication rounds) so as to adapt to the dynamics of the observed phenomena. Besides sensor networks, the issue of adaptive sampling has recently landed in the broader area of IoT systems and applications [33], again with the primary goal of optimising energy consumption of devices while not losing relevant phenomena under observation. However, in these contexts, such optimisations typically take place in a centralised (cloud) [32] or semi-decentralised (fog) way [18], which again disregards spatial issues and the strict space–time relations of phenomena.

6.2.2 Space-Aware Computing and Adaptive Spatial Sampling

Switching to considering the spatial dimension, we are interested in approaches that can adapt data processing to some spatial measure, for instance the density or extent of the phenomenon under observation—e.g. air pollution requiring denser sampling in regions with more variance and more sparse sampling in those with less variance. A few research works pursue similar objectives to the one we aim at with our notion of space fluidity.

Topology Adaptive Spatial Clustering [36] is a distributed algorithm that partitions a Wireless Sensor Network (WSN) into a set of disjoint sampling clusters, with no prior knowledge of cluster number or size (like ours), by encoding geographical distance, connectivity, and deployment density information in a single measure upon which leader election (for cluster heads) happens. The goal is to group together nodes in proximity and within regions of similar deployment density, to improve efficiency of data aggregation and compression. Besides the focus on efficiency, that is only half the story in our approach (the other being accuracy, hence the trade-off), two are the fundamental differences our approach has with respect to this: First, in our case, adaptation is domain-dependent, in the sense that it depends on some property (e.g. variance) of the phenomenon under observation, not on infrastructural properties; second, we do also consider over-sampling when variation is high, whereas reference [36] only considers under-sampling where measures are redundant.

Another distributed approach to adaptive spatial sampling [20] is based on the assumption that neighbouring sampling nodes usually have similar readings (i.e. high spatial correlation), hence can be grouped in a cluster to improve energy consumption. The proposed algorithm uses such spatial correlation, two application-specific threshold parameters (error tolerance and correlation range), and residual energy to elect cluster heads, with the goal of minimising the number of clusters, and the variance of their size. However, most of the calculations still rely on infrastructural properties rather than measures about the phenomenon of interest, and the focus is, once again, on energy saving; thus, authors do not consider over-sampling but solely under-sampling.

In [19], an approach is proposed that controls adaptation based on the information observed by sensors, rather than on network, energy, or other infrastructural aspects. However, they consider a special case where the clusters must correspond to predetermined “sources of interest”, such as physical objects/devices in the environment. Moreover, clusters are formed with the secondary objective of being balanced, whereas we allow them to be of different sizes and shapes (and even encourage to be so depending on the spatial distribution of the phenomenon of interest).

SILENCE [17] is a distributed, space–time adaptive sampling algorithm based on (space–time) correlation of measured values. Its main goal is efficiency: Indeed, SILENCE strives to reduce communication and processing overhead, by minimising sampling redundancy and reducing the scheduling speed of sensor devices. However, it focusses on the case where spatial correlation of sampled values is high.

Finally, the *ASample* algorithm [31] considers fully distributed domain-driven adaptation, also accounting the opportunity to over-sample, too. In particular, ASample builds a Voronoi tessellation of the area where the WSN is deployed, in a fully distributed way by considering only neighbourhood data. Such a tessellation promotes a desired application-level sampling accuracy: While a given Voronoi region is within the accuracy bound, it keeps expanding; on the contrary, whenever the accuracy constraint is violated, a virtual centroid of a novel Voronoi region is spawned, with a value that is obtained through interpolation of the neighbouring regions. This is a key aspect to consider, as it introduces synthetic data, which is something we avoid as we only increase sampling granularity if there are actual devices available in the target area. However, ASample assumes that the smaller the area covered by the Voronoi region, the less representative the samples drawn are, hence the smaller the impact on the global sampling accuracy. In our targeted scenarios, the opposite could be true, too: Smaller regions represent sharp variance of measurements across space and more accurately represent the irregularities of the underlying phenomenon. Indeed, in our proposal, we are fine that potentially irregular clusters reflect the irregular spatial distribution of the observed phenomenon.

6.3 Fluidity in Time

This section presents a model for time-fluid field-based coordination, emphasising the integration of a *causality field* within the field calculus framework [4]. The central concept involves utilising field-based coordination to govern the dynamics of computations in application-level fields.

Our proposed approach is broadly applicable to any field-based coordination framework, with a specific focus here on the general field calculus framework for clarity. The execution of field calculus programs yields *computational fields*, mapping *events* (space–time points representing computation rounds) to computational values. Inter-event communication forms a directed acyclic graph known as an *event structure* [22]—see Fig. 6.1 for a graphical example, modelling the interactions in a system of computing and interacting devices.

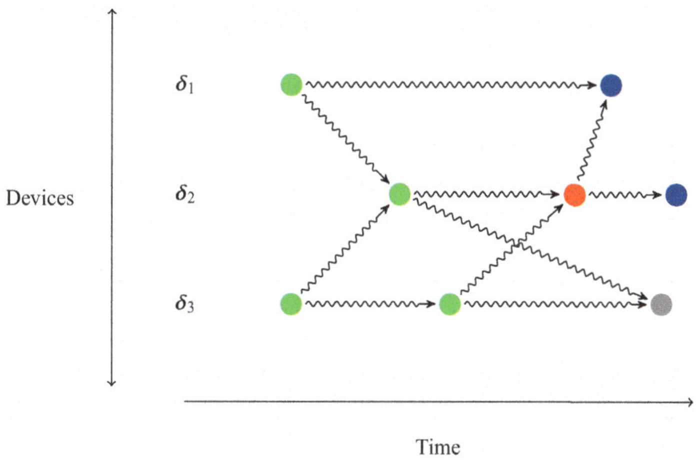

Fig. 6.1 Example of an event structure. Each node is an event (i.e. a computation round happening at a given device), and the curly arrows denote the causal relationship between events resulting from communication, which defines a partial order on the events. Given a reference event (red), through the $\rightsquigarrow$ relationship, it is possible to define its "causal past" event cone (green), "causal future" event cone (blue), and concurrent events (grey). Conventionally, we assume that the events horizontally aligned to the device labels δ_i are rounds performed by those devices. Along the arrows, information can be transferred, modelling state persistence in a given device or communication between different devices: In this example, δ_1 and δ_3 both communicate bidirectionally with δ_2. Moreover, since devices might be situated in space, and the event structure captures a distributed, platform-level notion of time, the events can denote space–time locations and, together with the corresponding computed values, a space–time computational field (modelling, e.g. a temperature field, a warning field, a field of suggestions for crowd dispersal, etc.)

Locally, events are scheduled by the middleware platform, serving as a low-level, asynchronous, distributed "clock". This local event scheduling provides the foundation for defining higher-level, time-fluid concepts. As a practical example, in a landslide monitoring application, events correspond to sensor readings assessing terrain movement, and the computational field represents the probability of a landslide occurring in the next T minutes.

6.3.1 A Time-Fluid Model

For a field calculus program **P**, each round involves processing valid messages from neighbours, $M \in \mathcal{M}$, and contextual information $S \in \mathcal{S}$ obtained via sensors. In our reference landslide monitoring scenario, S could include terrain movement measurements, and M messages from neighbouring sensors. The executing platform must decide whether to launch the next **P** evaluation round at a given event, providing valid values for $\mathcal{M}$ and $\mathcal{S}$. Multiple programs can be executed concurrently.

To support causality-driven coordination, we introduce a *(local event) trigger* associated with each event, representing the cause for the event. Triggers involve changes at the application level ($\mathcal{M}$) or physical level ($\mathcal{S}$), e.g. "new message arrived" or "sensor provides a new value". $\mathcal{T}$ denotes the set of possible local event triggers.

Additionally, we introduce a *guard policy*, denoted as G, acting as a scheduler for a "parent" field computation. A guard policy, expressed in the same language as **P**, is viewed locally as $f_G : (\mathcal{S}, \mathcal{M}) \rightarrow \mathcal{V}$. This allows multiple guard policies to be associated with a field computation, forming a *tree structure*. Platform triggers schedule low-level guard policies, forming the leafs, and higher-level policies use their outcomes to schedule subsequent evaluations.

Considering our landslide monitoring example, triggers may be sensor-related events, low-level guard policies could compute landslide probability based on sensor readings, and higher-level policies may visualise a landslide frontier. Feedback from a guard policy's evaluation can regulate its next evaluation, aiding in self-regulation. Refer to Fig. 6.2 for a graphical representation.

6.3.2 Causality Function and Temporal Dynamics

The scheduling impact of a guard policy is governed by the *causality function*, f_C, defined at the platform level. For a given round and program P, it takes the current round's trigger, the result of evaluating P at the previous round, and the results of its n lower-level schedulers, returning a Boolean. This Boolean denotes whether P is scheduled for evaluation at the current round, creating a *causality field* that acts as a self-regulated clock.

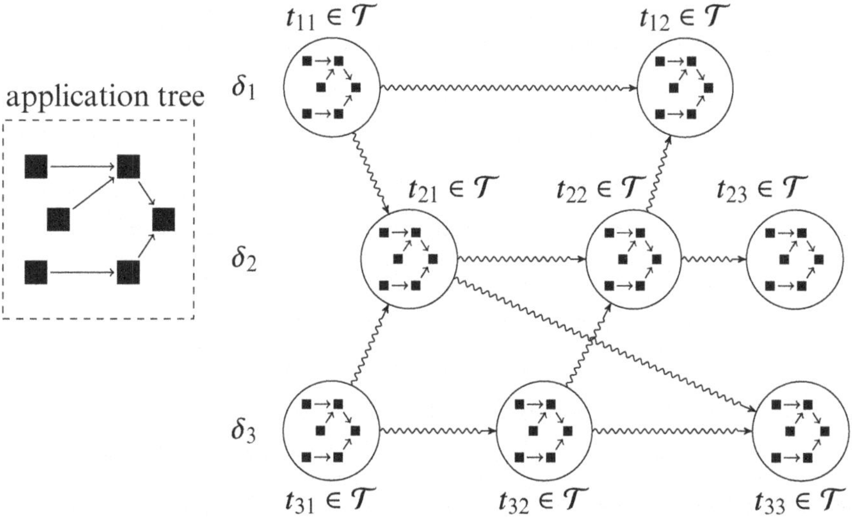

Fig. 6.2 This is an example of an event structure (structurally equivalent to that of Fig. 6.1) where each event (circle) is an evaluation of a tree of field calculus programs (the tree of black squares shown in the dashed box on the left). The labels on the nodes indicate the platform triggers causing the events to happen. The application tree has a root program (the right-most node) and schedulers as children and within its sub-trees; the nodes with no incoming arrows are scheduled merely on the basis of platform triggers

In this framework, a hierarchy of schedulers processes low-level triggers, culminating in a top-level causality field that schedules application-level computation. This structured guard facilitates precise temporal control.

The model's expressivity is influenced by the ability to sense context ($\mathcal{S}$) and express event triggers ($\mathcal{T}$). The platform can generate events due to triggers involving changes to any $\mathcal{S}$ value, making the computation reactive to device perception. Triggers flipping from `false` to `true` can model timers, encompassing the classic time-driven approach as a special case.

6.3.3 *Implications*

The introduced model has significant implications for the expressiveness of field-based coordination.

6.3.3.1 Programming Time and Propagating Causality

Enabling an application to influence its own execution policy allows programming time directly. Evaluating a field computation at different frequencies modulates the application's perception of time, impacting sensor sampling rates and communication dynamics across space. This consequence arises from adopting a causality-based notion of time, distinguishing this model from others.

6.3.3.2 Adapting to Causality

The model's capability to influence the space–time fabric requires awareness of the dynamics of causal relations among events. The application adapts not only to events' movements across time and space but also to changes in their "traveling speed". This dynamic adaptation enhances the system's ability to respond to varying environmental conditions.

6.3.3.3 Controlling Situatedness

Fine control over the degree of situatedness can be achieved by deciding the granularity of perceived event triggers and how to adapt to changes in event dynamics. This capability is crucial for rapidly reacting to dynamic environmental changes in distributed systems.

6.3.3.4 Co-causal Field Computation

Associating field computations with programmable scheduling policies allows expressive control but also introduces risks of circular dependencies. While offering increased expressiveness, circular dependencies must be carefully managed to avoid undesirable global behaviours like deadlocks or livelocks. The model implementation can and should be provided with static analysis and simulation tools to assess and mitigate such issues.

6.3.3.5 Pure Reactivity and Its Limitations

The model shifts the system from time-driven to event-driven, resulting in purely reactive behaviour. However, the inclusion of clock triggers and retroactive feedback mechanisms in guard policies provides flexibility and mitigates this limitation, allowing for proactive strategies in coordination policies. For instance, it is possible (and reasonable) to express a policy such as "trigger as soon as event ϵ happens, or timer τ expires, whichever comes first".

6.3.4 *Time-Fluid Aggregate Computing*

Building upon the previously discussed model, we introduce the notion of *time-fluid aggregate computing guards*. These guards represent a mechanism capable of dynamically adjusting the scheduling of field computations in response to environmental dynamics. The following sections are dedicated to presenting an extension of the field calculus semantics, specifically designed to integrate this model (Fig. 6.3). Additionally, we will detail an implementation of this model within the Protelis programming language, demonstrating the practical applicability and effectiveness of time-fluid aggregate computing guards in adaptive distributed systems.

6.3.5 Device-Wise Configuration and Semantics

This section formalise a time-fluid extension of the field calculus operational semantics (see Fig. 6.3), which is a distributed coordination model for field-based computations. A global field calculus scheduled program, denoted as $\mathsf{e}[\overline{\pi}]$, integrates a standard expression e with a sequence of scheduling programs $\overline{\pi}$. This structure forms a hierarchical tree where the root expression yields the field computation result, intermediate nodes represent scheduler expressions for their parent nodes, and leaf nodes are expressions scheduled directly by the platform, indicated as $\mathsf{e}[\bullet]$. The evaluation of these expressions involves exchanging an exported message μ structured as a value tree $\theta[\overline{\mu}]$, reflecting the tree's hierarchical nature and applying field calculus operational semantics node-wise.

The operational semantics for scheduling are encapsulated in the formula "$\delta; \Lambda; x \vdash \pi \Downarrow^{s} \mu$", interpreting the evaluation of a scheduling program π to an exported message μ on a device δ considering its local status field Λ and sensor state x. Here, Λ maps neighbouring devices to their messages, and x maps sensor names to their values.

Evaluation proceeds bottom-up within the tree at each round, contingent upon a scheduling context that includes the outcomes of lower-level scheduler programs, the previous round's evaluation result of the expression, and the trigger event information. Depending on whether the scheduling context activates the evaluation, the expression is either re-evaluated, updating the exported message with a new value tree, or the previous value tree is reused.

The formal difference between these scenarios is captured by the [ROUND] rule, which processes a program $\mathsf{e}[\overline{\pi}]$ against a neighbourhood. It recursively evaluates each scheduler π_i, generating a segment of the exported message μ_i, and then determines whether the main expression e should be evaluated based on the $\mathsf{sch}()$ predicate. This predicate abstractly determines the necessity of re-evaluation by considering the scheduling context, which includes the scheduler evaluations, the latest value tree of the device, and the trigger sensor's value.

a) Field calculus syntactic elements:

$\mathtt{v}$	field calculus value	$\mathtt{e}$	field calculus expression
θ	field calculus value-tree	δ	device unique identifier
$\mathtt{n}$	sensor name		

b) Field calculus abstracted semantics:

$\delta;\overline{\delta}\mapsto\overline{\theta};\overline{\mathtt{n}}\mapsto\overline{\mathtt{v}}\vdash \mathtt{e}_{\mathtt{main}}\Downarrow\theta$ — round semantics judgment

c) Local and scheduling configurations:

$\pi ::= \mathtt{e}[\overline{\pi}]$	scheduling program
$\mu ::= \theta[\overline{\mu}]$	exported message
$\Lambda ::= \overline{\delta}\mapsto\overline{\mu}$	local status field
$x ::= \overline{\mathtt{n}}\mapsto\overline{\mathtt{v}}$	local sensor state

d) Scheduling semantics: $\boxed{\delta;\Lambda;x\vdash\pi\Downarrow^{s}\mu}$

$$\text{[ROUND]}\ \frac{\begin{array}{c}\{\forall i,\ \delta;\overline{\delta}\mapsto\overline{\mu}^{i};x\vdash\pi_i\Downarrow^{s}\mu_i,\ \mu_i=\theta_i^{o}[\overline{\mu}^{0}]\}\\ \begin{cases}\theta=\theta_j & \text{if } \neg\mathtt{sch}(\theta_j,\overline{\theta}^{o},x(\mathtt{trigger}))\\ \delta_j;\overline{\delta}\mapsto\overline{\theta};x\vdash\mathtt{e}\Downarrow\theta & \text{otherwise}\end{cases}\end{array}}{\delta_j;\overline{\delta}\mapsto\overline{\theta}[\overline{\mu}^{1},\ldots,\overline{\mu}^{n}];x\vdash\mathtt{e}[\overline{\pi}]\Downarrow^{s}\theta[\overline{\mu}]}$$

e) System configurations:

$I ::= \overline{\delta}$	neighbourhood
$\tau ::= \overline{\delta}\mapsto\overline{I}$	topology field
$\Sigma ::= \overline{\delta}\mapsto\overline{x}$	sensor field
$\Psi ::= \overline{\delta}\mapsto\overline{\Lambda}$	status field
$Env ::= \tau:\Sigma$	environment
$N ::= \langle Env;\Psi\rangle$	network configuration

f) Environment well-formedness:

$WFE(\tau:\Sigma)$ holds iff $\mathbf{dom}(\tau)=\mathbf{dom}(\Sigma)$ and $\{\delta\}\subseteq\tau(\delta)\subseteq\mathbf{dom}(\Sigma),\forall\delta\in\mathbf{dom}(\Sigma)$

g) Network semantics: $\boxed{N\xrightarrow{act}N}$

$$\text{[N-FIR]}\ \frac{\begin{array}{c}Env=\tau:\Sigma \qquad \delta;\Psi(\delta);\Sigma(\delta),\mathtt{trigger}\mapsto\mathtt{v}_t\vdash\pi_{\mathtt{main}}\Downarrow^{s}\mu\\ \Psi_1=\tau(\delta)\mapsto\{\delta\mapsto\mu\}\end{array}}{\langle Env;\Psi\rangle\xrightarrow{\delta:\mathtt{v}_t}\langle Env;\Psi[\Psi_1]\rangle}$$

$$\text{[N-ENV]}\ \frac{WFN(Env') \qquad Env'=\overline{\delta}\mapsto\overline{I}:\overline{\delta}\mapsto\overline{x} \qquad \Psi_0=\overline{\delta}\mapsto\overline{I}\mapsto\mu_{\perp}^{\pi_{\mathtt{main}}}}{\langle Env;\Psi\rangle\xrightarrow{env}\langle Env';\Psi_0[\Psi]\rangle}$$

Fig. 6.3 Operational semantics for time-fluid coordination in the field calculus

6.3.6 System-Wise Configuration and Semantics

Building upon the scheduling semantics, the operational semantics for the distributed system's behaviour is derived, closely following the methodology presented in [4], adapted for scheduling formalisation. The system assumes the existence of a unique aggregate program $\pi_{\texttt{main}}$ and defines a network configuration N as a combination of an environment *Env* and a global status field Ψ. The environment comprises a topology field τ and a sensor state field Σ, mapping devices to their neighbours and sensor states, respectively, and Ψ represents the communication status across devices.

A network configuration is considered valid if the sensor and topology fields align in domain, with the topology being *reflexive* and *domain-closed*. The operational semantics ensure the preservation of environment well-formedness through transitions, defined by two rules: [N-FIR], modelling a scheduling round triggered by a device, and [N-ENV], representing environmental changes. The [N-FIR] rule updates the global status field to reflect the message exchange between neighbours post-evaluation, while [N-ENV] accounts for changes in topology or sensor states, ensuring the global status field is updated to maintain system integrity and message compliance.

6.3.7 Time-Fluid Scheduling in Protelis

The operational semantics presented in Fig. 6.3 need concrete implementations for certain abstracted components. These include:

1. Identification of triggers ($\mathtt{v}_t$ from sensor `trigger`) causing a local round via rule [N-FIR]
2. Definition of the scheduling tree, allowing the application developer to integrate a main script `e` with its dependent scheduling programs $\overline{\pi}$
3. Implementation and specification of the `sch()` predicate to manage scheduling logic based on the aforementioned scheduling structure

In our model, scripts form a tree structure of program nodes, each representing a named Protelis module. Dependencies between modules ($\pi \rightarrow \pi'$) denote program node relationships, mapping a program node to π (equivalent to $\mathtt{e}[\overline{\pi}]$), with Protelis modules corresponding to `e` and dependencies to $\overline{\pi}$. Execution scheduling follows the operational semantics, requiring only the specification of the Protelis module at the tree's root to schedule the entire tree. Execution precedence is given to modules without dependencies, followed by those whose dependencies have been executed, culminating with the root module.

Execution consideration involves evaluating the $\mathtt{sch}(\theta, \overline{\theta}^{o}, x(\mathtt{trigger}))$ predicate for each module, a task delegated to the platform. Implementation can vary, such as a unified function for the entire tree or individual functions per module. We adopt the latter, introducing an API for Protelis modules to define their scheduling predicates via a `scheduler` function, as demonstrated in Listing 1. This function, invoked with parameters injected via annotations, determines module execution.

Annotations `@Input` and `@Changed` are utilised for binding sensor values and detecting value changes, respectively, facilitating local scheduling logic without aggregate constructs. Static analysis of Protelis programs referenced by the main script enables automatic inference of the program tree from source code without additional configuration.

```
module a:b
def scheduler(
  @Input("trigger") t,
  @Input("a:b") old,
  @Input("c:d") someSchedulerOutput,
  @Changed("c:d") didChange
) = // local scheduling logic

// function declarations
// main expression
```

Listing 1 Implementation of the sch(θ, $\overline{\theta}^{o}$, x(`trigger`)) predicate per module

6.3.8 Examples of Time-Fluid Guards

This section presents examples that leverage our scheduling approach for time-fluid aggregate computations. We focus on scenarios where platform-provided triggers (i.e. sensor `trigger` values) play a crucial role. These triggers, highlighted in blue, often lead to scheduler functions acting as logical disjunctions (ORs) of input conditions or simple predicates, aligning with the intuition that a computation should proceed whenever *any* of its potential triggers occur.

6.3.8.1 Timer-Based Scheduling

This example showcases a policy recreating the classic, timer-based execution model, thus demonstrating how this approach subsumes the original execution model of field computations [4]. Consider a chain of events (and the corresponding triggers) local to one device such as the following.

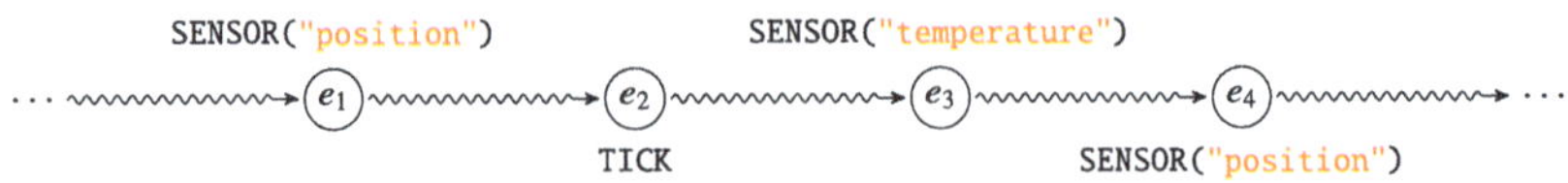

Assuming the platform exposes a sensor for obtaining the current local timestamp in seconds, it is possible to define a timer-like scheduler through a program as in Listing 2. There is no declared scheduler, so it runs in every event. Its output

is a Boolean field mapping a Boolean value to each event. We use notation ⇓ in the following picture to indicate the input (portion) and output of an event corresponding to a certain program execution.

```
module timefluid:every_second

// Applies condition to current and to the last value that made condition
// true, returning the truth value and storing the new state if needed
def updated(init, current, condition) =
  rep(state <- [init, true]) {
    let old = state.first()
    if (condition(current, old)) {
      [current, true]
    } else {
      [old, false]
    }
  }.get(1)

updated(-1, self.getCurrentTime()) { now, last -> now-last>=1 }
```

Listing 2 Timer-like scheduler

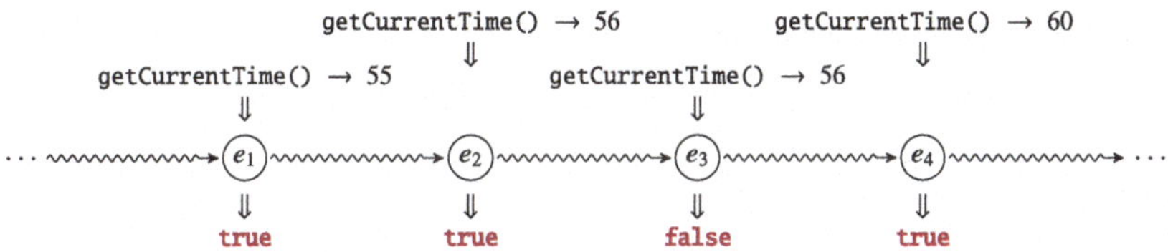

Notice that you could not actually schedule "every second" if the underlying platform events are not triggered with a second or sub-second frequency; moreover, here we only generally consider soft real-time tasks. So, for instance, a downstream program that needs to run at most once per second will not be scheduled in round e_3. Such a program can be defined with a scheduling predicate logic that merely reuses the output of module `timefluid:every_second` as in Listing 3. Function `scheduler` is used by the platform to control the scheduling of `some_program`: It is invoked with the set of triggers of the current event and other input parameters corresponding to the upstream scheduling programs.

```
module timefluid:some_program

// Special function 'scheduler' maps input scheduling fields to Boolean
def scheduler(@Input("timefluid:every_second") new_t) = new_t

// main expression
```

Listing 3 A scheduling predicate that reuses `timefluid:every_second`

An alternative would be to define a base timer trigger EVERY_SECOND at the platform level that is issued every second.

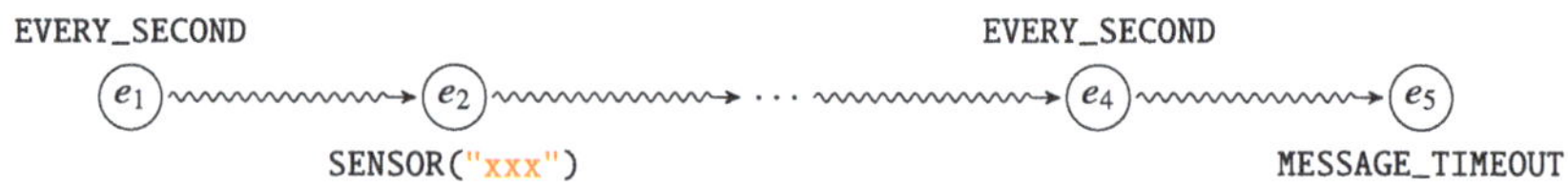

Once we have a user-defined program or platform-level timer, derived timers could be defined as in Listing 4.

```
module timefluid:every_minute

import platform.Triggers.EVERY_SECOND

// OPTION A) restricting the domain of a high-level scheduler
def scheduler(@Input("timefluid:every_second") s) = s
// OPTION B) restricting the domain of a low-level, platform
    scheduler
def scheduler(@Input("trigger") t) = t == EVERY_SECOND

// returns true once every 60 timer ticks
rep(old <- -1) { old + 1 }  60 == 0
```

Listing 4 Example timers derived from domain restriction

Reacting to Local Context Changes

On one end of the policy spectrum, we have a *purely* reactive execution model. Here, local computations are triggered only under specific conditions: a change in sensor values (SENSOR(".*")), the receipt of new information (MESSAGE_RECEIVED), or the expiration of information from a neighbour (MESSAGE_TIMEOUT). An example of such a policy is shown below: This approach ensures computations are only performed when necessary, based on sensor data or message flow, promoting efficiency and responsiveness.

```
module timefluid:some-program

import platform.Triggers.*

def scheduler(@Input("trigger") t) =
  t == SENSOR(".*") ||
  t == MESSAGE_RECEIVED("timefluid:some-program") ||
  t == MESSAGE_TIMEOUT("timefluid:some-program")
```

Listing 5 A purely reactive scheduling policy

Leveraging Aggregate Computations for Scheduling

In scenarios like crowd monitoring at large events, two main tasks emerge: estimating crowd density and steering the crowd to prevent congestion. Efficient management requires updating steering directives only upon significant changes in crowd density.

We achieve this by employing the SCR pattern to partition space and calculate average crowd density within 300-metre regions, as demonstrated in the Protelis program Listing 6.

```
module timefluid:steering:density
import ...

def scheduler(@Input("trigger") t) =
  t == SENSOR("people_count") ||
  t == MESSAGE_RECEIVED("timefluid:steering:density") ||
  t == MESSAGE_TIMEOUT("timefluid:steering:density")

let distToLeader = distanceTo(S(300))
let count = summarize(distToLeader, env.get("people_count"), sum)
let radius = summarize(distToLeader, distToLeader, max)
count / (2 * PI * radius)
```

Listing 6 Estimating local crowd density using the SCR pattern

The density computation's result then triggers another computation for steering, as shown next, which is scheduled based on changes in the density estimate (Listing 7). Finally, the steering logic itself is predicated on the output of the scheduler, forming a causal chain of computations that adaptively respond to crowd dynamics (Listing 8), also depicted in Fig. 6.4.

```
module timefluid:steering:steering-scheduler

def scheduler(@Changed("timefluid:steering:density") d) = d

changed(exponentialBackOff(env.get("timefluid:steering:density"),
     0.1))
{
  cur, old -> abs(cur - old) > 0.5
}
```

Listing 7 Scheduling based on changes in crowd density

```
module timefluid:steering:steering

def scheduler(@Input("timefluid:steering:steering-scheduler") sched)
     = sched
```

Listing 8 Crowd-steering program scheduled by density changes

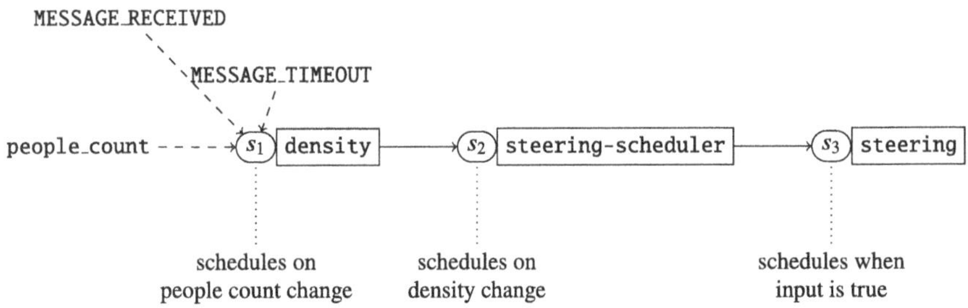

Fig. 6.4 This figure illustrates how Protelis modules, represented as square nodes, and their schedulers, shown as rounded rectangles, interact through triggers and data flows, depicted by solid and dashed arrows, respectively. This design enables dynamic and efficient crowd management by closely linking computation to changes in the environment

6.4 Fluidity in Space

In this section, we devise distributed coordination strategies for the estimation of a spatial phenomenon through collaborative adaptive sampling. Our design is based on the idea of dynamically partitioning space into regions that compete and grow/shrink to provide sampling preserving the required accuracy.

6.4.1 Problem Definition

We start by introducing the notion of *regional partition*, which is a finite set of non-overlapping contiguous clusters of devices: a notion that prepares the ground to that of an *aggregate sampling* which we introduce in this chapter.

Definition 6.1 (Regional Partition Field, Contiguous Regions) Let $\mathbf{E} = \langle E, \rightsquigarrow, d\rangle$ be a stabilising environment static since event ϵ_0. A *regional partition field* is a stabilising field $f : E \to \mathbb{V}$ on $\mathbf{E}$ such that:

- **(Finiteness)** The image $Img(f) = \{f(x) \mid x \in E\}$ is a finite set of values.
- **(Eventual contiguity)** There exists an event $\epsilon_0' > \epsilon_0$ such that for any pair of events $\epsilon_1, \epsilon_n \in T^+_{\epsilon_0'}$, $f(\epsilon_1) = f(\epsilon_n)$ implies that there is a sequence of events $\epsilon_1 < \ldots < \epsilon_n$ connecting ϵ_1 to ϵ_n where $f(\epsilon_i) = f(\epsilon_1) = f(\epsilon_n)\ \forall 1 \leq i \leq n$.

Note that the set of domains of regions induced by f is defined by *regions*$(f) = \{f^{-1}(v) : v \in Img(f)\}$. Moreover, given two regions $E, E' \in$ *regions*(f), we say that they are *contiguous* if $\exists \epsilon \in E, \epsilon' \in E' : \epsilon \rightsquigarrow \epsilon' \vee \epsilon' \rightsquigarrow \epsilon$.

An example of a regional partition field is shown in Fig. 6.5. Notice that for any pair of events in the same space–time region there exists a path of events entirely contained in that region. Also, notice that, by this definition, different disjoint regions denoted by the same value r are not possible.

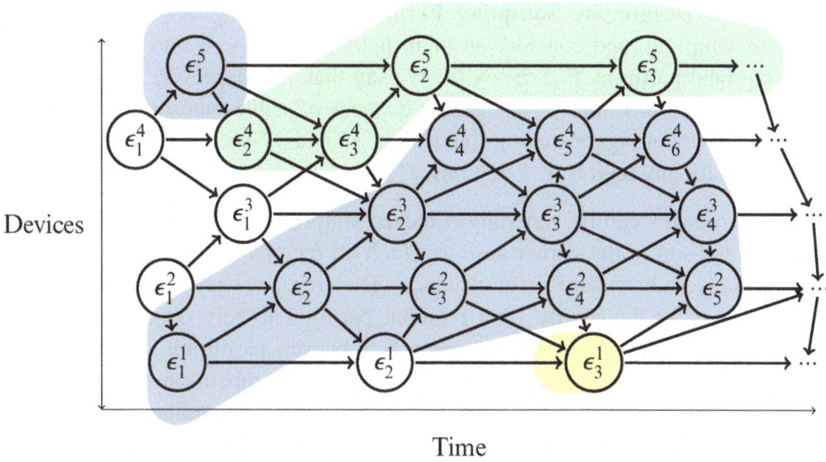

Fig. 6.5 Example of a regional partition field with regions r_{blue}, r_{green}, r_{yellow}, r_{white} (the background is used to denote the output of the field). Notice that contiguity does not hold everywhere and anytime but only since event ϵ_2^3

Definition 6.2 (Aggregate Sampling) An *aggregate sampling* is a stabilising computation $\Phi_S : \mathbb{F}_{E,\mathbb{V}} \rightarrow \mathbb{F}_{E,\mathbb{V}}$ that, given an input field to be sampled, yields as output a regional partition field.

Once we have defined an aggregate sampling process in terms of its inputs, outputs, and stabilising dynamics, we need a way to measure the error introduced by the aggregate sampling. To this purpose, we introduce the notion of a *stable snapshot*, namely a field consisting of a sample of one event per device from the stable portion of a stabilising field.

Definition 6.3 (Stable Snapshot) Let $\mathbf{E} = \langle E, \rightsquigarrow, d\rangle$ be an event structure and $f : E \rightarrow \mathbb{V}$ be a stabilising field on $\mathbf{E}$ which provides stable output from $\epsilon_0 \in E$. We define a *stable snapshot* of field f as a field obtained by restricting f to a subset of events in the future event cone of ϵ_0 *and* with exactly one event per device, i.e. a field $f_S : E_S \rightarrow \mathbb{V}$ such that $E_S \subseteq T^+_{\epsilon_0}$, *and* $\forall \epsilon, \epsilon' \in E_S : d(\epsilon) = d(\epsilon') \implies \epsilon = \epsilon'$, *and* $\forall \epsilon \in T^+_{\epsilon_0}, \exists \epsilon' \in E_S : d(\epsilon') = d(\epsilon)$.

Definition 6.4 (Stable Snapshot Error Distance) We call *stable snapshot error distance* any metric $\mu : \mathbb{F}_{E,\mathbb{V}} \times \mathbb{F}_{E,\mathbb{V}} \rightarrow \mathbb{R}_0^+$ over stable snapshots that feature same domain (event structure) and codomain (a set of values).

We are now able to characterise adequacy properties for a sampling operator, intuitively capturing the fact that sampling correctly trades off the size of regions with their accuracy. We first start by introducing a notion that handles accuracy, stating that any of the produced regions will not cause the error distance to be over a certain threshold.

Definition 6.5 (Aggregate Sampling Error) Let $\Phi_{\mathbf{E}} : \mathbb{F}_{E,\mathbb{V}} \to \mathbb{F}_{E,\mathbb{V}}$ be an aggregate sampling, and consider an input field $f_i : E \to \mathbb{V}$ and corresponding output regional partition $f_o : E \to \mathbb{V}$. We say that f_o *samples* f_i *within error* η *according to error distance* μ, if the error distance of stable snapshots of f_i and f_o in any region is not bigger than η, that is: Let f_i^s and f_o^s be stable snapshots of f_i and f_o; then for any region $E' \in regions(f_o^s)$, we have $\mu(f_i^s|_{E'}, f_o^s|_{E'}) \leq \eta$.

Note that *accuracy* can be generally achieved simply by partitions defining many small regions—up to the corner case in which all regions include just one device, hence trivially induce zero error distance. Therefore, we are also interested in *efficiency*, namely the ability of a regional partition to rely on as few regions as possible. Without a centralised approach, however, partitioning is necessarily suboptimal, since it can rely only on local interaction/competition among regions; hence it should be expected that some regions will stop "expanding" as they reach a smaller threshold. Additionally, there can also be corner cases where regions with very small error distance are created, e.g. because what remains to be covered in an iterative selection of regions is simply a very small part of the network, or one with rather uniform values introducing little sampling errors. What we may require from an adequate sampling operator, however, is that such regions are somewhat not the norm. This is formally captured by the following definition, essentially introducing a "lower bound" for the error distance of regions.

Definition 6.6 (Local Optimality of a Regional Partition) Let $\Phi_{\mathbf{E}} : \mathbb{F}_{E,\mathbb{V}} \to \mathbb{F}_{E,\mathbb{V}}$ be an aggregate sampling, consider an input field f_i and corresponding output regional partition f_o such that f_o samples f_i within error η according to distance μ, and denote with f_i^s and f_o^s the stable snapshots of f_i and f_o, respectively (cf. Definition 6.5). We say that f_o is *locally optimal* under error η and with efficiency k ($k > 0$) if all pairs of contiguous regions $E', E'' \in regions(f_o)$ are such that $\mu(f_i^s|_{E' \cup E''}, f_o^s|_{E' \cup E''}) \geq k \cdot \eta$.

For example, we will show that the algorithm we propose guarantees $k = 0.5$ (see Sect. 6.4.5). Note that we call this notion "local optimality" to stress the fact that an identified partition is not necessarily the best one that could be found, but it is one that cannot be significantly improved with a small change, such as combining two regions—a small improvement is possible, depending on the efficiency factor k. This notion well fits our goal of dealing with dynamic phenomena and large-scale environments, where one is more geared towards finding good heuristics for self-organising behaviour.

So, we are now ready to define the goal operator.

Definition 6.7 (Effective Sampling Operator) An *effective sampling operator* with efficiency k is a self-stabilising operator P_η, parametric in the error bound η, such that in any stabilising environment $\mathbf{E}$ and stabilising input f_i, a locally optimal regional partition with efficiency k and within error η is produced.

6.4.2 Aggregate Computing-Based Solution

In this section, we define a *space-based adaptive sampling* algorithm, called AGGREGATESAMPLER, discuss its implementation in *aggregate computing* [5, 35], and prove the algorithm is a self-stabilising, effective sampler with efficiency at least $k = 0.5$.

6.4.3 AGGREGATESAMPLER *Algorithm for Adaptive Spatial Sampling*

The problem of creating partitions in a self-organising way is very much related to a problem of multi-leader election [25, 27]. Building on this idea, our approach starts by solving a *sparse leaders election problem* [21], for which self-stabilising solutions exist [8, 23, 25].

Leaders are used as *samplers* of the input field. During the election of leaders/samplers, we associate them with larger and larger regions of "follower devices" that will provide the sampled value as output. During execution of the algorithm, such regions will expand until the desired error distance can be kept under the threshold η. This process is managed so that there will not be any overlap with other regions, and so that no devices of the network remain outside of some region (i.e. each device will follow exactly one leader).

To ensure that regions are connected and will not overcome the threshold independently of the chosen leader, we adopt as error distance one based on "distance among devices", as follows. The algorithm can be configured to adopt any strategy that is able to turn input and output fields into a metric m for devices: Such a metric is as usual a function mapping a pair of neighbour devices to a non-negative real number, called the "local sampling distance" of the two devices—intuitively, the higher the physical distance of devices and the higher the difference of input values and output values of the two devices, the higher is m for that pair. Given this metric, any pair of devices of the network can be associated with a *path sampling error*, which is the size of the shortest path (according to the metric) connecting the two devices. The proposed algorithm will then produce regions adopting as error distance μ the maximum path sampling error of any pair of devices in the region, and it will turn out that any pair of contiguous regions combined will necessarily give error distance greater than $0.5 * \eta$ (efficiency $k = 0.5$).

The algorithm is defined as follows (see Fig. 6.6):

1. Each device announces its candidature for leadership.
2. Each device propagates to its neighbours the candidature of the device it currently recognises as leader, its sampled value, and the path sampling error from it, fostering the expansion of its corresponding region.

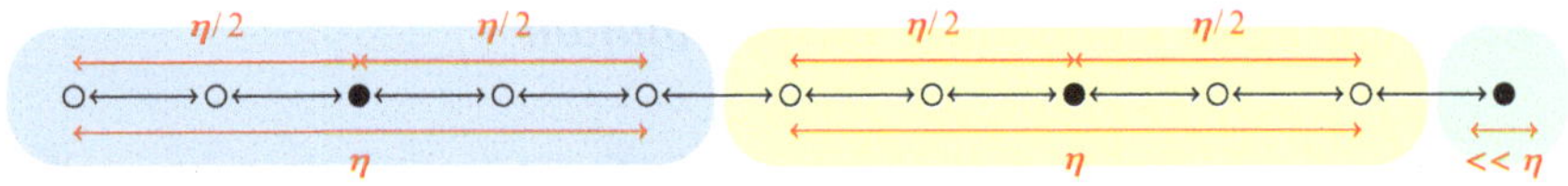

Fig. 6.6 Example of a regional partitioning (with three contiguous regions) created by the algorithm on a simplified system where devices are arranged on a line. Notation: Black dots denote the leaders/samplers; the coloured areas denote regions; and the red extension lines are used to denote the error distances. Note that no device in a region can have path sampling error greater than $\eta/2$ with respect to the leader and that very small regions can still exist in corner cases (e.g. the green region on the right)

3. Devices discard candidatures whose path sampling error from the leader exceeds half the expected threshold ($\eta/2$).
4. In case multiple valid candidatures (i.e. those that are not discarded) reach a device, one is selected based on a *competition policy*.

The specific strategies for computing the local sampling distance and the leader competition policy are application-dependent.

6.4.3.1 Competition and Leader Strength

Although competition among leaders could be realised in several ways, many techniques may lead to non-self-stabilising behaviour: For instance, if the winning leader is selected randomly in the set of those whose error is under threshold, regions may keep changing even in a static environment. In this chapter, we propose a simple strategy: Every leader associates its candidature with the local value of a field that we call *leader strength*; in case of competing candidatures, the highest such value is selected as winner, breaking the symmetry. The leader strength can be of any orderable type, and its choice impacts the overall selection of the regions by imposing a selection priority over leaders (hence on region-generation points). If two candidate leaders have the same strength, then we prefer the closest one. If we are in the (unlikely) situation of perfect symmetry, with two equally strong candidate leaders at the same distance, then their device identifier is used to break symmetry.

6.4.3.2 Region Expansion and Path Sampling Error

Inspired by previous work on distributed systems whose computation is independent of device distribution [6], the proposed approach essentially accumulates the path sampling error along the path from the leader device towards other devices along a gradient, a distributed data structure that can be generated through self-stabilising computations [34]. We thus have two major drivers:

1. The leader strength affects the creation of regions by influencing the positions of their source points.
2. The path sampling error influences the expansion in space of the region across all directions, mandating its size and (along with the interaction with other regions) its shape.

For instance, a metric could be the absolute value of the difference in the perceived signal (e.g. a value sampled from a sensor) between two devices: Devices perceiving very different values would tend not to cluster together (even if spatially close), as they would perceive each other as farther away (leading to irregular shapes).

6.4.4 Aggregate Computing-Based Implementation

An implementation of the algorithm expressed in the Protelis aggregate programming language [28] is shown in Fig. 6.7. In aggregate computing, a so-called *aggregate program*, such as the one shown in Fig. 6.7, is repeatedly run by all the devices: It expresses a logic for mapping the *local context* (given by sensor readings and messages from neighbours) to an *output value* and an *output message* to be sent to all the neighbours. In other words, in aggregate computing, each event denotes a full execution of the aggregate program against the event's inputs, determining the message payload passed to receiving events.

The core of the program is function AGGREGATESAMPLER, which consists of the following main elements:

- *Sampler candidacies are modelled as ordered triplets*, using 3-element tuples, with corresponding accessor functions (Fig. 6.7, Lines 1–6), of the elements:
  1. `simmetryBreaker`: a value used to break symmetry, capturing the "strength" of a candidacy
  2. `distance`: a value capturing the distance to the sampler node of a candidacy
  3. `leaderId`: holding the device identifier of the candidate sampler
- Function `expansionLogic` (Fig. 6.7, 8–13) is defined to determine when a candidacy has to be discarded, i.e. when it comes from the device itself or when the distance of the candidate sampler is greater than or equal to a `radius` parameter.
- Function AGGREGATESAMPLER (Fig. 6.7, 15–24) is the entry point of the algorithm, parametrised in terms of the executing device identifier (`mid`), a maximum spatial range of candidacies (`radius`), a value to break symmetry (`symmetryBreaker`), and a `metric` function providing distances to neighbours.
- `share(x <- init){ e }` is a bidirectional communication construct [3] that works as follows: The declared variable `x`, which is set to `init` at the first round, collects the evaluations of the overall `share` expression in neighbour devices (including the device itself), and the new value for the current device (which is the data item that will be sent to neighbours) is obtained by evaluating expression `e`.

```
1  // Definition of the record (product type) Sample + accessor functions
2  def Sample(symmetryBreaker, distance, leaderId) = [symmetryBreaker,
       distance, leaderId]
3  def breakSymmetry(sample) = sample.get(0)
4  def sampleDistance(sample) = sample.get(1)
5  def areaCenter(sample) = sample.get(2)
6  def discard() = Sample(POSITIVE_INFINITY, POSITIVE_INFINITY,
       POSITIVE_INFINITY)
7  // Logic to control the propagation of candidacies
8  def expansionLogic(sample, localId, radius) =
9    mux (areaCenter(sample) == localId || sampleDistance(sample) >= radius
       ) {
10     discard()
11   } else {
12     sample
13   }
14
15 def AggregateSampler(mid, radius, symmetryBreaker, metric) {
16   let local = Sample(-symmetryBreaker, 0, mid)
17   areaCenter(
18     share (received <- local) {
19       let candidacies = received.set(1, received.get(1) + metric())
20       let filtered = expansionLogic(candidacies, mid, radius)
21       min(local, foldMin(discard(), filtered))
22     }
23   )
24 }
25
```

Fig. 6.7 Source code of the algorithm

- *The distance field of the neighbour candidacies is updated* (Fig. 6.7, Line 19), by adding to each candidacy provided by a neighbour the local distance w.r.t. that neighbour[1] (as provided by `metric`).
- *Neighbours' candidacies that are too far or support the current device are deprioritised* (Fig. 6.7, Lines 20), by function `expansionLogic()`.
- Finally, *selecting the winner over the processed candidacies through minimisation*, by `min` and `foldMin` (Fig. 6.7, Line 21), which minimise over the filtered candidacy triplets, with default candidacy as the one with the lowest priority (provided by `discard()`), and against the `local` candidacy.

The algorithm described above is an effective sampling operator (Definition 6.7) as long as (i) *half the path sampling error η is used as parameter* `radius`, so that function `expansionLogic` does not expand regions beyond the given error η, and (ii) *an additive* `metric` *is used*, so that it is impossible to decrease the error by expanding any given region (at best, it will stay the same).

[1] Note that this operation, together with the `share` application, essentially provides the same structure as the basic gradient algorithm.

6.4.5 *Formal Analysis*

In this section, we prove that the proposed solution is self-stabilising and that it represents an effective sampling operator leading to a bounded-error locally optimal regional partition (cf. Definitions 6.2, 6.6, 6.7).

Since our aggregate sampling must be a stabilising computation (see Definition 6.2), we start by proving that our algorithm is self-stabilising. We do so by exploiting the framework in [3, 34], which defines a set of self-stabilising *fragments* that can be composed together to yield self-stabilising operators. In particular, in [34] it is proved that any closed expression in the self-stabilising fragment is self-stabilising, by structural induction on the syntax of expressions and programs ([34], Appendix E, Lemma 2): Values and variables are already self-stabilised, a function application self-stabilises (by the inductive hypothesis) if its arguments are self-stabilising, and similar considerations can be done for the other program fragments.

Theorem 1 (AGGREGATESAMPLER **Is Self-Stabilising**)

Proof In [3], it is proved that an expression of the following form (called a *minimising share* pattern) is self-stabilising:

```
share(x <- e) { fR(minHoodLoc(fMP(x), e), x) }
```

where (see Section 5.2 in [34] and, especially, Section 4.7 and Figure 7 in [3]):

- `fR(x, prev)` is a "raising function", with respect to partial orders, of `x` and `prev` (the value of `x` at the previous round).
- `fMP` is a monotonic progressive function of `x`, which can take further arguments as far as they are self-stabilising expressions that do not contain the `share`-bounded variable `x`.
- `minHoodLoc(e, loc)` selects the minimum among the neighbours' values of expression `e` and the current device's local value `loc`.

Now, the block of Protelis code (Fig. 6.7) in Lines 15–24 can be rewritten as shown in Fig. 6.8 that conforms to the minimising `share` pattern, where:

- The raising function `fR` is an identity on the first parameter, which is a trivially valid raising function (see Example 5.5 in [34]).
- Function `expansionLogic` is a valid monotonic progressive function `fMP` of `x`, since it transforms neighbours' candidacies supporting the current device and those at a distance farther than `radius` to the highest value for the data type (cf. `discard()`), leaving the others unaltered; none of the provided additional arguments (`id`, `radius`, and `metric`) contains the `share`-bounded variable `x`.

More gradually, the transformation can be obtained by:

1. Renaming `received` to `x`, and defining functions `fR`, `fMP`, and `minHoodLoc`.
2. Realising that `fMP` is a valid replacement for the combination of distance field update and call to `expansionLogic` and replacing accordingly.

```
def updateDistance(x, metric) {
  x.set(1, x.get(1) + metric())
  x
}
def fR(x, prev) = x // raising function
def fMP(x, localId, radius, metric) = // monotonic progr.
  expansionLogic(updateDistance(x, metric), localId, radius)
def minHoodLoc(e, loc) = // minimum of loc and e's values
  min(loc, foldMin(discard(), e))
def AggregateSampler(mid, radius, symmetryBreaker, metric) {
  let local = Sample(-symmetryBreaker, 0, mid)
  share (x <- local) {
    fR(minHoodLoc(fMP(x, mid, radius, metric), local), x)
  }.get(2)
}

```

Fig. 6.8 Protelis code from Fig. 6.7 rewritten to conform to the *minimising share* self-stabilising pattern as per the proof

3. Replacing `areaCenter` with its definition.
4. Replacing the `share` body with `minHoodLoc`, as they perform the same operation.
5. Adding a call to `fR`, which is an identity function, does not alter the behaviour of the code and leads directly to the code in Fig. 6.8.

The other elements in the program are only operations on local data which are also self-stabilising expressions. Since the AGGREGATESAMPLER function consists exclusively of self-stabilising expressions, it is in turn self-stabilising [3, 34].

Theorem 2 (**AGGREGATESAMPLER Is an Effective Sampling Operator**)

Proof To prove that our algorithm represents an effective sampling operator P_η = AGGREGATESAMPLER, we have to prove that it yields, on any stabilising input field f_i, an output stabilising field f_o of locally optimal regional partitions. As per 1, P_η is self-stabilising, so on a stable input it will yield a stable output: Let f_o^s be its snapshot, and f_i^s the corresponding input.

On the one hand, accuracy is guaranteed since AGGREGATESAMPLER ensures that no device has path sampling error greater than $\eta/2$ from the leader: For the triangular inequality property of metric spaces ($m(a,b) \leq m(a,c) + m(c,b)$) this ensures that stable snapshot distance μ does not overcome η.

On the other hand, for local optimality under error η and distance μ, there must not exist two contiguous regions $E', E'' \in \mathit{regions}(f_o^s)$, with samplers δ' and δ'', where $\mu(f_i^s|_{E' \cup E''}, f_o^s|_{E' \cup E''}) \leq \eta/2$. Suppose two such regions exist, and let δ' be stronger than δ'' (i.e. it has higher symmetry breaker). Then, the path sampling error between δ' and δ'' is necessarily higher than $\eta/2$, because of the steps 3 and 4 of the algorithm: In fact, if it were smaller than $\eta/2$, then δ'' would have followed δ' and would not have been a leader (cf. Fig. 6.9).

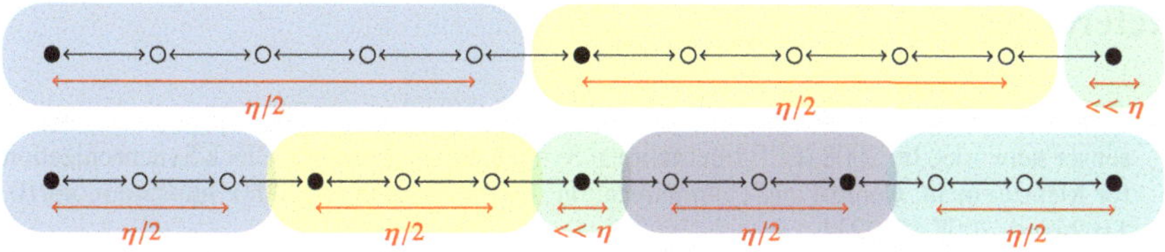

Fig. 6.9 Examples of regional partitionings with efficiency 0.5. Notation: Black dots denote the leaders; the coloured areas denote regions; and the red extension lines are used to denote the error distances. Note how the union of the green singleton region (associated to the weakest leader) with its neighbouring regions would make the error distance of the latter exceed $\eta/2$—if that would not be the case, then the former region would not have existed in the first place (cf. Theorem 2)

6.5 Conclusion

In this chapter we presented two works presenting a first operational definition of, respectively, time-fluid and space-fluid distributed computations, modelled and implemented in the conceptual and technological framework offered by the field calculus. These works should give both an intuition and a first formal account of how the kind of fluid computations envisioned by FluidWare can be actually enabled, supported, and leveraged by currently available models and platforms.

About time-fluid computations, a natural extension of such model is to consider learning mechanisms to adapt the scheduling of field computations [2] as discussed in chapter "Learning Opportunities in Collective Adaptive Systems" of this book. Another direction lies in the investigation of programming paradigms especially tailored to handling causality and propagation of change, such as functional reactive approaches [9].

Conversely, a natural next step along these lines would be to integrate the two proposed approaches into a unique time–space-fluid computing platform, where computations can seamlessly adapt to the temporal evolution of the phenomena they deal with, as well as to their spatial dynamics. The fact that both approaches have been both theoretically and technically grounded in the field calculus formal model and programming toolchain can greatly help in this endeavour. However, care must be used regarding two concerns: First of all, there is an overhead in supporting the described adaptation mechanisms directly at the platform (i.e. middleware level) that must be measured and taken into account when considering applications with some sort of timing and performance constraints; second, the quest towards autonomous adaptation that we are pursuing must not let control slip away from our hands, as system engineers must have a way to intervene on these platform-level adaptation mechanisms when due and "override" their emergent behaviour with the desired one.

We are well aware of these concerns and hope that the proposed first step in the direction of time- and space-fluid computations that we have taken can pave the way for others to explore similar research avenues.

References

1. Ageev, A., Macii, D., Flammini, A.: Towards an adaptive synchronization policy for wireless sensor networks. In: 2008 IEEE International Symposium on Precision Clock Synchronization for Measurement, Control and Communication. IEEE, Piscataway (2008). https://doi.org/10.1109/ispcs.2008.4659224
2. Aguzzi, G., Casadei, R., Viroli, M.: Addressing collective computations efficiency: towards a platform-level reinforcement learning approach. In: Casadei, R., Nitto, E.D., Gerostathopoulos, I., Pianini, D., Dusparic, I., Wood, T., Nelson, P.R., Pournaras, E., Bencomo, N., Götz, S., Krupitzer, C., Raibulet, C. (eds.) IEEE International Conference on Autonomic Computing and Self-Organizing Systems, ACSOS 2022, Virtual, CA, USA, September 19–23, 2022. pp. 11–20. IEEE, Piscataway (2022). https://doi.org/10.1109/ACSOS55765.2022.00019
3. Audrito, G., Beal, J., Damiani, F., Pianini, D., Viroli, M.: Field-based coordination with the share operator. Log. Methods Comput. Sci. **16**(4) (2020). https://lmcs.episciences.org/6816
4. Audrito, G., Viroli, M., Damiani, F., Pianini, D., Beal, J.: A higher-order calculus of computational fields. ACM Trans. Comput. Logic **20**(1), 1–55 (2019). https://doi.org/10.1145/3285956
5. Beal, J., Pianini, D., Viroli, M.: Aggregate programming for the internet of things. IEEE Comput. **48**(9), 22–30 (2015). https://doi.org/10.1109/MC.2015.261
6. Beal, J., Viroli, M., Pianini, D., Damiani, F.: Self-adaptation to device distribution in the internet of things. ACM Trans. Auton. Adapt. Syst. **12**(3), 12:1–12:29 (2017). https://doi.org/10.1145/3105758
7. Bicocchi, N., Mamei, M., Zambonelli, F.: Self-organizing virtual macro sensors. ACM Trans. Auton. Adapt. Syst. **7**(1), 2:1–2:28 (2012). https://doi.org/10.1145/2168260.2168262
8. Burman, J., Chen, H., Chen, H., Doty, D., Nowak, T., Severson, E.E., Xu, C.: Time-optimal self-stabilizing leader election in population protocols. In: PODC '21: ACM Symposium on Principles of Distributed Computing, Virtual Event, Italy, July 26–30, 2021, pp. 33–44. ACM, New York (2021). https://doi.org/10.1145/3465084.3467898
9. Casadei, R., Dente, F., Aguzzi, G., Pianini, D., Viroli, M.: Self-organisation programming: a functional reactive macro approach. In: IEEE International Conference on Autonomic Computing and Self-Organizing Systems, ACSOS 2023, Toronto, ON, Canada, September 25–29, 2023, pp. 87–96. IEEE, Piscataway (2023). https://doi.org/10.1109/ACSOS58161.2023.00026
10. Casadei, R., Mariani, S., Pianini, D., Viroli, M., Zambonelli, F.: Space-fluid adaptive sampling by self-organisation. Log. Methods Comput. Sci. 19(4) (2023). https://doi.org/10.46298/lmcs-19(4:29)2023
11. Fischer, M.J.: The consensus problem in unreliable distributed systems (a brief survey). In: International Conference on Fundamentals of Computation Theory, pp. 127–140. Springer, Berlin (1983)
12. Gershenson, C.: Design and control of self-organizing systems. CopIt Arxives (2007)
13. Hélary, J.M., Hurfin, M., Mostefaoui, A., Raynal, M., Tronel, F.: Computing global functions in asynchronous distributed systems with perfect failure detectors. IEEE Trans. Parallel Distrib. Syst. **11**(9), 897–909 (2000)
14. Ho, Y., Huang, Y., Chu, H., Chen, L.: Adaptive sensing scheme using naive Bayes classification for environment monitoring with drone. Int. J. Distrib. Sensor Netw. **14**(1) (2018). https://doi.org/10.1177/1550147718756036
15. Jelasity, M., Montresor, A., Babaoglu, Ö.: Gossip-based aggregation in large dynamic networks. ACM Trans. Comput. Syst. **23**(3), 219–252 (2005). https://doi.org/10.1145/1082469.1082470
16. Kho, J., Rogers, A., Jennings, N.R.: Decentralized control of adaptive sampling in wireless sensor networks. ACM Trans. Sensor Netw. **5**(3), 19:1–19:35 (2009). https://doi.org/10.1145/1525856.1525857

17. Lee, E.K., Viswanathan, H., Pompili, D.: SILENCE: distributed adaptive sampling for sensor-based autonomic systems. In: Proceedings of the 8th International Conference on Autonomic Computing, ICAC 2011, Karlsruhe, Germany, June 14–18, 2011, pp. 61–70. ACM, New York (2011). https://doi.org/10.1145/1998582.1998594
18. Lee, J., Yoon, G., Choi, H.: Monitoring of IoT data for reducing network traffic. In: Tenth International Conference on Ubiquitous and Future Networks, ICUFN 2018, Prague, Czech Republic, July 3–6, 2018, pp. 395–397 (2018). https://doi.org/10.1109/ICUFN.2018.8436601
19. Lin, Y., Megerian, S.: Sensing driven clustering for monitoring and control applications. In: 4th IEEE Consumer Communications and Networking Conference, CCNC 2007, Las Vegas, NV, USA, January 11–13, 2007, pp. 202–206. IEEE, Piscataway (2007). https://doi.org/10.1109/CCNC.2007.47
20. Liu, Z., Xing, W., Zeng, B., Wang, Y., Lu, D.: Distributed spatial correlation-based clustering for approximate data collection in WSNs. In: 27th IEEE International Conference on Advanced Information Networking and Applications, AINA 2013, Barcelona, Spain, March 25–28, 2013, pp. 56–63. IEEE Computer Society (2013). https://doi.org/10.1109/AINA.2013.26
21. Lynch, N.A.: Distributed Algorithms. Morgan Kaufmann (1996)
22. Mattern, F., et al.: Virtual time and global states of distributed systems. University, Department of Computer Science (1988)
23. Mo, Y., Audrito, G., Dasgupta, S., Beal, J.: A resilient leader election algorithm using aggregate computing blocks. IFAC-PapersOnLine **53**(2), 3336–3341 (2020). https://doi.org/10.1016/j.ifacol.2020.12.1497
24. Parunak, H.V.D., Brueckner, S.A.: Software engineering for self-organizing systems. Knowl. Eng. Rev. **30**(4), 419–434 (2015). https://doi.org/10.1017/S0269888915000089
25. Pianini, D., Casadei, R., Viroli, M.: Self-stabilising priority-based multi-leader election and network partitioning. In: IEEE International Conference on Autonomic Computing and Self-Organizing Systems, ACSOS 2022, Virtual, CA, USA, September 19–23, 2022, pp. 81–90. IEEE, Piscataway (2022). https://doi.org/10.1109/ACSOS55765.2022.00026
26. Pianini, D., Casadei, R., Viroli, M., Mariani, S., Zambonelli, F.: Time-fluid field-based coordination through programmable distributed schedulers. Log. Methods Comput. Sci. **17**(4) (2021). https://doi.org/10.46298/lmcs-17(4:13)2021
27. Pianini, D., Casadei, R., Viroli, M., Natali, A.: Partitioned integration and coordination via the self-organising coordination regions pattern. Fut. Gener. Comput. Syst. **114**, 44–68 (2021). https://doi.org/10.1016/j.future.2020.07.032
28. Pianini, D., Viroli, M., Beal, J.: Protelis: practical aggregate programming. In: Proceedings of the 30th Annual ACM Symposium on Applied Computing, Salamanca, Spain, April 13–17, 2015, pp. 1846–1853 (2015). https://doi.org/10.1145/2695664.2695913
29. Singh, V.K., Singh, G., Pande, S.: Emergence, self-organization and collective intelligence—modeling the dynamics of complex collectives in social and organizational settings. In: UKSim, pp. 182–189. IEEE, Piscataway (2013). https://doi.org/10.1109/UKSim.2013.77
30. Sundararaman, B., Buy, U., Kshemkalyani, A.D.: Clock synchronization for wireless sensor networks: a survey. Ad Hoc Netw. **3**(3), 281–323 (2005). https://doi.org/10.1016/j.adhoc.2005.01.002
31. Szczytowski, P., Khelil, A., Suri, N.: Asample: adaptive spatial sampling in wireless sensor networks. In: IEEE International Conference on Sensor Networks, Ubiquitous, and Trustworthy Computing, SUTC 2010 and IEEE International Workshop on Ubiquitous and Mobile Computing, UMC 2010, 7–9 June 2010, Newport Beach, California, USA, pp. 35–42. IEEE Computer Society (2010). https://doi.org/10.1109/SUTC.2010.37
32. Traub, J., Breß, S., Rabl, T., Katsifodimos, A., Markl, V.: Optimized on-demand data streaming from sensor nodes. In: Proceedings of the 2017 Symposium on Cloud Computing, SoCC 2017, Santa Clara, CA, USA, September 24–27, 2017, pp. 586–597 (2017). https://doi.org/10.1145/3127479.3131621
33. Trihinas, D., Pallis, G., Dikaiakos, M.: Low-cost adaptive monitoring techniques for the internet of things. IEEE Trans. Serv. Comput. (2018). https://doi.org/10.1109/tsc.2018.2808956

34. Viroli, M., Audrito, G., Beal, J., Damiani, F., Pianini, D.: Engineering resilient collective adaptive systems by self-stabilisation. ACM Trans. Model. Comput. Simul. **28**(2), 1–28 (2018). https://doi.org/10.1145/3177774
35. Viroli, M., Beal, J., Damiani, F., Audrito, G., Casadei, R., Pianini, D.: From distributed coordination to field calculus and aggregate computing. J. Log. Algebraic Methods Program. **109**, 100486 (2019). https://doi.org/10.1016/j.jlamp.2019.100486
36. Virrankoski, R., Savvides, A.: TASC: topology adaptive spatial clustering for sensor networks. In: IEEE 2nd International Conference on Mobile Adhoc and Sensor Systems, MASS 2005, November 7–10, 2005, The City Center Hotel, Washington, USA, p. 10. IEEE Computer Society (2005). https://doi.org/10.1109/MAHSS.2005.1542850

Part III
Visions

Chapter 7
Fluidware Meets Digital Twins

Roberto Casadei, Fabrizio Fornari, Stefano Mariani, and Claudio Savaglio

7.1 Introduction

A rapid rise of interest in the potential of Digital Twins (DTs) to transform a vast range of Internet of Things (IoT) and cyber-physical system (CPS) applications [37] has recently emerged.

A DT is a virtual representation of a physical object, process, or system. It uses real-time data and simulations to mimic the behaviour, characteristics, and interactions of its real-world counterpart [29, 30]. The concept of a DT allows organizations to gain insights into the performance, maintenance, and optimization of physical assets and processes. It brings together various technologies, such as the Internet of Things (IoT), data analytics, artificial intelligence (AI), and cloud computing, to create a comprehensive and dynamic representation of the physical entity in the digital realm. The rapid advancements in the mentioned digital technologies have paved the way for the development and utilization of DTs that allow bridging the gap between physical systems and their virtual representations.

R. Casadei
Università degli Studi di Bologna, Cesena, Italy
e-mail: roby.casadei@unibo.it

F. Fornari
Università degli Studi di Camerino, Camerino, Italy
e-mail: fabrizio.fornari@unicam.it

S. Mariani (✉)
Università di Modena e Reggio Emilia, Reggio Emilia, Italy
e-mail: stefano.mariani@unimore.it

C. Savaglio
Università della Calabria, Rende, Italy
e-mail: csavaglio@dimes.unical.it

F. Zambonelli et al. (eds.), *Fluidware*, Internet of Things,
https://doi.org/10.1007/978-3-031-62146-8_7

In particular, the Internet of Things is seen as the key enabler to filling the gap between the physical and virtual worlds, and it has been defined as the backbone for the development and use of Digital Twins [33]. The value that DTs add to the management, development, and commercialization of IoT systems can have a disruptive impact in the whole ICT landscape of the next few years, further bridging the physical–virtual divide. The 2023 Gartner emerging tech impact radar places DTs among the most impactful emerging technologies and trends [61].

The idea of twinning physical assets is over 50 years old, but the true rise of DTs corresponded with the IoT. Indeed, simultaneously, on one hand the plethora of IoT solutions has enabled main DT functionalities, solving those technological and infrastructural issues that have limited DTs spread so far; on the other hand, the wide sea of heterogeneous IoT devices, data, and services requires adequate abstraction, encapsulation, prototyping, management, and commercialization solutions that can be successfully pivoted on the DT concept. In fact, implementing DT solutions is not a trivial task. Industry pioneers and researchers interpreted the DT concept during the last years, each reporting and emphasizing different characteristics [35, 57]. As a result, no recognized standard definition and no unified solution to design and implement all the characteristics and functionalities of a DT is available [3, 18]. Within the scope of the Fluidware project [26, 64], conceptual and practical solutions for the integration between DT and IoT have been explored as well as the state of the art of main DT-aided IoT Platforms reviewed [18, 27].

In this chapter we first provide in Sect. 7.2 an overview of the Digital Twin concept, of its characteristics, its enabling technologies, and application scenarios. We then report in Sect. 7.3 the contributions resulting from the Fluidware project that can be adopted in the process of developing Digital Twins solutions. In Sect. 7.4 we report a discussion on Digital Twins and the challenges that they face. Section 7.5 concludes the chapter with a short summary, envisioned implications, and future work.

7.2 Digital Twins Overview

In this section, we provide an overview on the characteristics of DTs, their enabling technologies, and the various application domains that can benefit from their adoption.

7.2.1 Digital Twin Characteristics

DTs are considered as digital models that represent and have a direct connection with real entities (e.g. through IoT sensors and actuators). Data collected directly from the physical entities are allowed to update their digital representations. From such digital representations, information can be extracted that may impact decision-

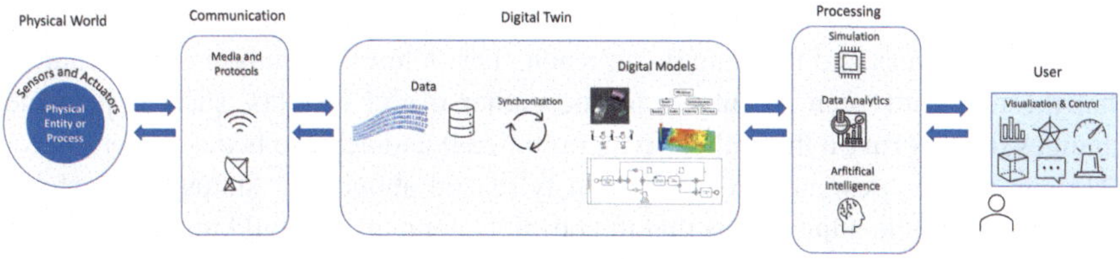

Fig. 7.1 Digital Twin concept overview

making and guide changes to be reflected in the real environment. Simulation techniques and AI adoption enable to predict failures, to optimize the system, to design novel features, to ease and accelerate decision-making, and to improve productivity. A DT can persist for the whole entity's life cycle, from design to implementation, deployment, maintenance, and disposal [31] (Fig. 7.1).

DTs can be used to reflect the appearance of a real entity, through the design and use of *graphical representations* such as 2D/3D models. Some models that characterize DTs are used to represent *states* in which a physical entity can be (e.g. turned on, turned off, active, idle, or error state), *behaviours* the physical entity can have, and the *context* in which the physical entity acts. These models are updated through data captured from the physical entity and environment, in such a way to properly monitor it and to reason about the actions and the status of the physical entity and the activity it is performing so to plan further operations. A physical entity can be composed of smaller components (e.g. machinery can have several mechanical parts), if necessary, each single component could have a respective digital representation and DT, leading to a *composition* of DTs.

As a DT is a digital representation of a physical entity, it is fundamental that a connection between the physical and the virtual replica exists so that the DT can receive and send data from/to the physical entity. This *interaction* is necessary to maintain a synchronization *between* the *physical and virtual entities* in such a way as to reflect changes from the virtual world to the physical one and vice versa. This process is often referred to as *twinning process* [35]. For example, an event occurring in the physical world will require updating the virtual replica to reflect changes to the digital entity, vice versa when an action is requested to a DT, a command could be sent to the physical counterpart so that it also performs that action. DTs can *exchange data with information systems* to collaborate and provide more elaborated services. This way, a DT can send and receive data from/to external sources or services. The possibility for DTs to *interact between each other* and to exchange data is seen as an important characteristic necessary for establishing networks or communities of DTs. In this way, DTs can exchange information and modify their behaviour according to the context they perceive.

DTs can process and analyse both historical and real-time data to understand the occurred events and the current status the real entity is in. DTs can rely on data analysis techniques to estimate when a certain situation will occur, e.g. when maintenance should be performed. DTs, as recognized by all the analysed literature, are used to conduct *simulations* of the activities the real entity can perform, or of the data it can process and the state it can reach. This allows to analyse case scenarios that the real entity can be subject to and gather useful insights, such as possible changes to perform on the real entity to optimize it and achieve better performance. DTs can also be designed to autonomously reason about their status and perform possible automatic adjustments that the physical counterpart could not do by itself.

7.2.2 Enabling Technologies

Digital Twin is certainly one of the paradigms (e.g. software agents, model-based engineering, grid computing) which has discovered newfound lifeblood and application fields with the emergence of IoT [11, 12, 28]. IoT data, technology, and protocols have thus empowered the DTs and fostered their evolution with the contribution of other relevant paradigms like AI, big data, etc. Indeed, the most advanced examples of DTs, also called Cognitive Twins or Intelligent DTs, are defined just in the context of IoT Platforms and leveraging on their connectivity, analytics, simulation, and visualization tools. Conversely, the value DTs can bring to the IoT is also relevant, with notable impacts on the design, prototyping, and commercialization of current and future IoT devices and services. These aspects are key since they typically hinder developers and organizations in fully achieving the IoT vision despite its inherent heterogeneity, large scale, and complexity.

In August 2020, within the context of the Object Management Group, the Digital Twin Consortium has started working on a taxonomy and on standards for DT enabling technologies, including the IoT. In particular, IoT-related technologies like EPC and RFiD can greatly support *DT identification*; ontologies coming from the IoT realm, like SensorML, SSN, and OWL, can also ease the *DT modelling*; XML, SOAP, JSON, and Flatbuffers have recently found success also for the *DT serialization* for the versatility and efficiency shown as IoT data interchange formats and management mechanisms; the plethora of network/message protocols paved by IoT (foremost LoRaWAN, Bluetooth, 802.11.*, ZigBee, MQTT, CoAP) can be exploited to support *DT connectivity* according to the specific requirements in terms of bandwidth and energy consumption, latency, overhead, coverage, data rate, wired–wireless transmission, etc.; if provided with IoT API (e.g. OGC Sensor-Things and Webinos IoT), DTs can access sensors and actuators in a transparent and secure way, thus achieving truly *cyber-physicality*; finally, microservices and containers are possible candidates for programmatically supporting the integration and composition of several IoT devices but also of DTs, automating their *deployment* and simplifying the design of their complex services.

These IoT-enabled DT functionalities are concretely implemented within the IoT Platforms software instruments generally used to organize and handle: IoT devices and their interactions; the data they capture from the environment; the manipulation, the processing, the visualization of such data; and the interaction with external service [12]. These fully support stakeholders in the scalable, efficient, and secure management (i.e. connection, access, protection, analysis, visualization) of heterogeneous (IoT, but not exclusively) data and devices, aiming to streamline development processes and provide greater value to business. To this end, an increasing number of IoT Platforms has started offering support for the creation, integration, and exploitation of DTs along their entire life (from the initial DT model tuning to the periodic synchronization given by the entanglement with the physical assets), leveraging on the rich data, analytics tools, and connectivity mechanism provided by the IoT Platform itself [34, 48] making them become suitable options for implementing DT solutions [4, 8, 53].

The landscape of IoT platforms is wider every day so, in the following, we briefly introduce ten of the most relevant ones provided by both cloud and industrial vendors: (i) Amazon Web Services (AWS); (ii) Azure Digital Twin; (iii) Bosch IoT Suite; (iv) Google Cloud IoT Core; (v) Lumanda; (vi) Predix; (vii) MindSphere; (viii) SAP Cloud Platform; (ix) ThingWorx ; and (x) Watson IoT Platform. In [18] a comparison of DT characteristics and IoT Platform functionalities is reported.

7.2.3 Digital Twin Application Scenarios

The DT paradigm is adopted nowadays in multiple application scenarios, ranging from Industry 4.0 [62], there including and especially manufacturing, to healthcare [41], and smart cities [58], there including smart mobility and intelligent transportation [52]. In the following we report two examples: one from the healthcare sector and the other focused on cooperative driving.

In healthcare, major trauma management is one of the challenging scenarios where DT-based solutions have been already conceived. There, physicians must have strong heterogeneous expertise, promptly identify a diagnosis, and quickly provide medical aid. In [54], a DT-based solution is outlined, conceptually structured in three stages:

1. Management of the emergency calls. For the trauma event, a new DT is created and linked to the DTs of both the available ambulance and the designated rescuer.
2. Pre-hospital management. DTs of patients are created and linked to the event DT, as well as to the trauma centre DT in charge of patient intake.
3. In-hospital management. The trauma event DT keeps tracking everything that happens to the patient and gets linked to the DTs of the team of physicians involved in trauma management, and to the hospital rooms needed.

All these DTs and especially their semantic interlink are used to support various kinds of agent-based applications, e.g. a personal assistant agent for the medical personnel, and a trauma tracking applications for documentation purposes.

DTs can have a major role in supporting cooperative driving with both autonomous and non-autonomous vehicles sharing the road infrastructure. There, basic services such as intersection crossing, parking, and ride sharing will need to be redesigned [45]. A WoDT can play the role of the enabling coordination infrastructure in the case of intersection crossing by following three major steps:

1. The municipality deploys the intelligent road infrastructure on a target intersection. Relevant Road Side Units (RSUs) are deployed, such as smart cameras to monitor traffic conditions. The DTs of the intersection and of the crossing process are created.
2. As soon as vehicles approach the intersection, their DTs are linked to the intersection and to the crossing process, and the coordination process to establish who gets the right of way.
3. Depending on the specific intersection crossing policy [45], the crossing process DT assigns the right of way or a waiting command to the vehicles' DTs, until all vehicles are served.

The whole crossing process is highly dynamic: Links among DTs are created whenever vehicles approach and leave the intersection, and DTs are used to continuously reflect (and control, wherever possible) the state of the involved assets, such as vehicles themselves, RSUs, as well as the (coordinated) crossing process.

For further scenarios, the interested readers can refer to surveys available in the literature [49].

7.3 Fluidware for Digital Twins

The Fluidware project led to several contributions that can be adopted to design and develop DT applications.

7.3.1 Designing Digital Twin Solutions

Fluidware also contributed with the development of MDE approaches combined with Low Code development for supporting the design and implementation of IoT application [17, 19–21, 24]. Such approaches support the design of IoT applications by means of graphical notations that abstract from the specificity of a single scenario. They allow handling the high variability of the IoT domain and enable the reuse of knowledge in the production of different solutions for IoT applications. Such successful approaches can also be applied in the design and development of DT solutions. A first contribution towards this direction involves the proposal of a graphical notation (DTMN—Digital Twin Modelling Notation) and software instruments that have been defined for supporting the design and conceptualization of DT solutions [18, 22]. A meta-model has been designed to incorporate the main characteristics of DTs, derived from a literature study; therefore the defined graphical notations allow to reason in terms of the physical *entity* or *system* that needs to be digitally represented; the *state* in which an entity can be and the

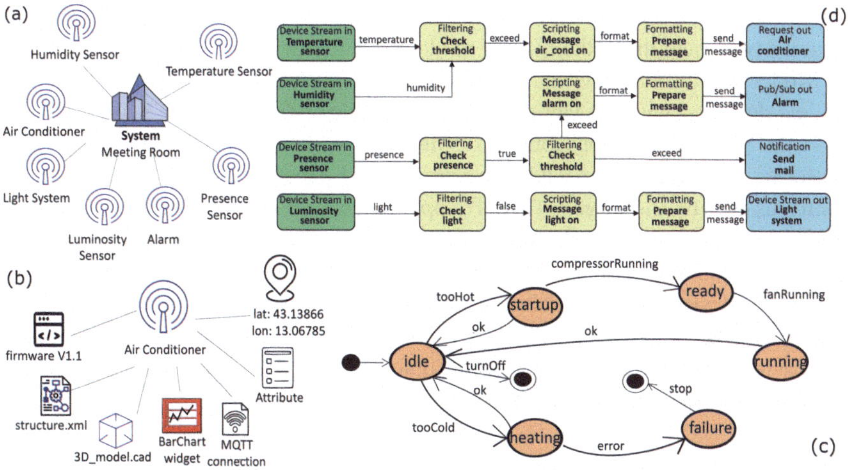

Fig. 7.2 Models representing (**a**) all the entities involved in a meeting room system, (**b**) a detailed visualization of an air conditioner Digital Twin, (**c**) the air conditioner state machine, and (**d**) the entire meeting room system application logic [18]

transitions that can cause state changes; the *visualization* of data and of the digital representation of the entity; *context* in which the physical twin is located; *data* that can be handled; and *application* that can be defined.

The usage of DTMN to design a DT solution for a sample scenario regarding a smart meeting room is reported in Fig. 7.2. Figure 7.2a illustrates the Context viewpoint. It represents a meeting room model along with the various entities involved. System and entity names, as well as descriptions, can be ascribed. Further specialized data relevant to each entity can also be integrated. Figure 7.2b provides details about a specific entity, in this case, an air conditioner. This entity could be intricately elaborated with supplementary components that align with the attributes of a DT, encompassing features like its composition, traits, spatial placement, connectivity, and associated descriptive files. Figure 7.2c reports the air conditioner's state machine perspective. Each DT can be outfitted with a state machine configuration, delineating the distinct states it can assume and the transitions that precipitate state alterations. Figure 7.2d reports the portion of the notation used to describe data manipulation activities and the overall application logic of a DT solution. It can be seen as a way to describe the management and control of the intelligent meeting room including: temperature- and humidity-based air conditioner regulation, dispatching notifications to the manager via email, triggering an alert in situations of room overcrowding, and the automatic management of lighting, driven by luminosity sensors.

In **Book Chapter A BPMN-Based Approach for IoT Systems Engineering**, we presented an MDE approach that starting from a domain-agnostic and standard notation called BPMN—Business Process Model and Notation [50]—allows describing collaborations between IoT processes carried on by multiple devices up to their execution made possible by the so-called BPMN engines [16]. Similarly to what we presented there, the DTMN modelling notation could lead to the definition of Model-Driven Engineering (MDE) approaches that can be used to go from models designed with DTMN to actual executable code. Given the strong relation between DT and IoT concepts, DT solutions are being developed through IoT platforms that have been extended to represent DT aspects [27, 53]. Such DT-ready platforms can be the target of MDE approaches, and code derived from models can be executed on such platforms. Especially, given the variety of IoT and DT platforms that are available in the market, the use of a cross-platform approach that allows the design of a DT solution once and the deployment on multiple target platforms, like proposed in [21], can be envisaged as a possible direction. This approach would allow DT solution designers to select multiple target platforms based on their specific support for each DT characteristic.

7.3.2 *Agentified Digital Twins*

Within the scope of the Fluidware project, efforts to take the lessons learned from the World Wide Web, the IoT, multiagent systems, and distributed systems, have also been conducted, and this led to the definition of an event-driven, decentralized, interoperable, linkable, and discoverable vision of DTs named Web of Digital Twins (WoDT) [47, 54]. The WoDT can be actually shaped into specific application domains with peculiar challenges and constraints related, for example, to the enrolment of heterogeneous physical assets, a structured hierarchical organization, and dynamic evolution in terms of interactions and knowledge representation. On the one hand, the WoDT allows to model DT's properties, behaviours, and relationships and consequently to represent large-scale and complex physical environments as an open ecosystem of connected and interoperable DTs (Fig. 7.3).

Also the potential synergies between agents and DTs, and multiagent systems (MAS) and (networks of) DTs, to reason about both the individual and collective (system) levels have been investigated. The proposed model, reported in Fig. 7.4,

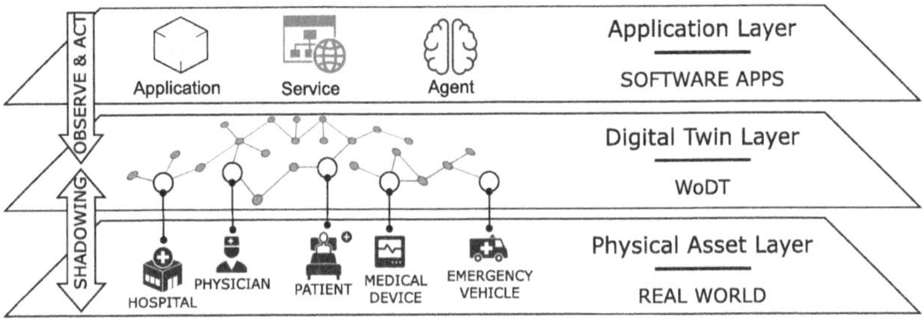

Fig. 7.3 WoDT layered view [54]

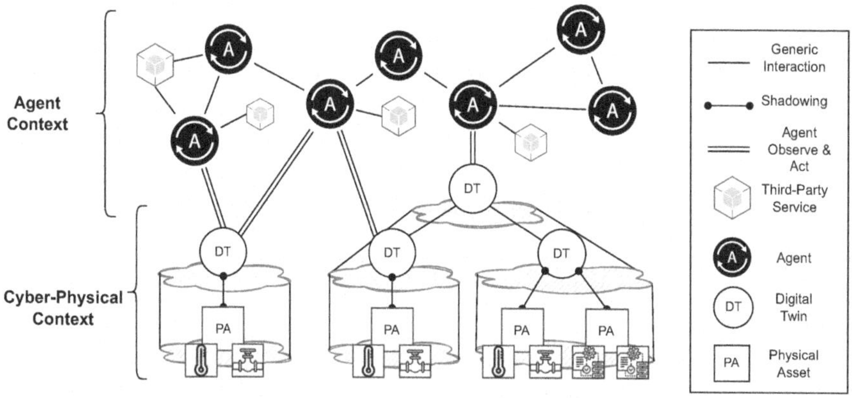

Fig. 7.4 Agent and Digital Twins relation [47]

and the abstract architecture define a basic conceptual framework that can be mapped onto a variety of concrete deployment scenarios and implementation technologies, with the aim to be a unifying horizontal layer on top of the physical assets. DTs are seen as entities interlinked in a web of semantic, dynamic relationships that enable structuring a dynamic application domain.

The proposed WoDT vision supports the definition of a new cyber layer where applications, agents, and services can implement and orchestrate new smart and dynamic systems of components by relying on a structured and integrated DT's overlay, without the responsibility to handle the fragmentation and the heterogeneity characterizing the physical layer. Moving forward from the local scope of a single application domain, the possibility to exploit a uniform and interoperable Web of DTs also opens the way to the design of a new generation of cross-domain computational infrastructures, trying to mirror the physical world where existing assets seamlessly move and interact across multiple contexts at the same time. For example, a person can be an employee for a company and a patient for the health system, or an ambulance can be a vehicle on the street and a resource for the trauma management ecosystem. Through the adoption of WoDT, DTs from multiple realms can start cooperating (potentially on demand) to reach a shared goal or to opportunistically implement a new behaviour that is something quite difficult to achieve in the siloed environments representing the state of the art.

7.3.3 Collective and Augmented Digital Twins

The engineering of self-organizing cyber-physical systems can benefit from DTs, virtual devices, and (augmented) collective DTs. In particular, collective DTs provide for a design construct towards collective computing, which can be augmented with virtual devices to improve the performance of existing self-organizing applications—as shown through swarm exploration and navigation scenarios [14].

Especially, the *collective computing* [2], or the *artificial collective intelligence* paradigm [13], can contribute to and integrate with the DT approach in two main ways [14]. First, a digital thread providing the bidirectional synchronization between a DT and the corresponding physical twin may be *supported by a collective process*, whereby multiple devices collaborate in order to provide an effective representation or reification of the target state. Secondly, considering collectives as first-class citizens in the design of socio-technical or multiagent systems suggests the possibility of *considering DTs for entire collectives* (e.g. the DT for a whole swarm of robots).

Collective computing techniques can be adopted to support the fidelity or the implementation of digital twins (DTs). For instance, the shape information about a physical entity may be supported in a DT by multiple cameras [42]. As another example, the mean temperature of a digitalized room may be computed by combining the contributions of multiple thermometers. Related *sensor fusion* techniques have been applied to DTs, e.g. in the smart vehicle domain [42].

A *Collective Digital Twin (CDT)* is a DT that has a collective nature, namely one that is associated to an entire collective [15]. Consider the DT of an entire *swarm* of robots: It can model swarm-level information that goes beyond the properties and state of an individual robot—e.g. macroscopic properties like the direction in which the swarm as a whole is moving, or results of consensus. A CDT may be implemented or represented as a collective of collaborating DTs (see Fig. 7.5). The definition of a CDT is a matter of reification and is motivated by the need of capturing macroscopic information as well as providing an abstracted monitoring and control interface over ensembles of physical entities.

The CDT notion can be *augmented* considering virtual devices that have no physical counterparts [14, 15]. This leads to a notion of *augmented CDT* (see Fig. 7.6), which provides the software model of an augmented reality system and enables activities like the steering of self-organizing behaviours [32, 38]. In [14, 15], for instance, virtual devices are spawned to increase the coverage of a spatial area in order to improve crowd-aware navigation and exploration processes. To implement this, two main problems have to be addressed: the creation of virtual

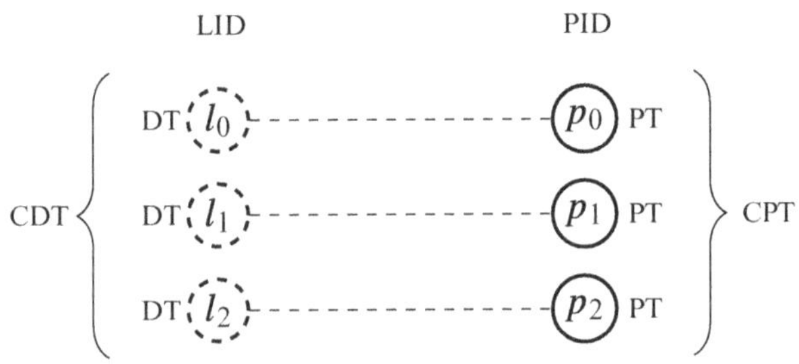

Fig. 7.5 Collectives of digital (CDT) and physical twins (CPT) [14]

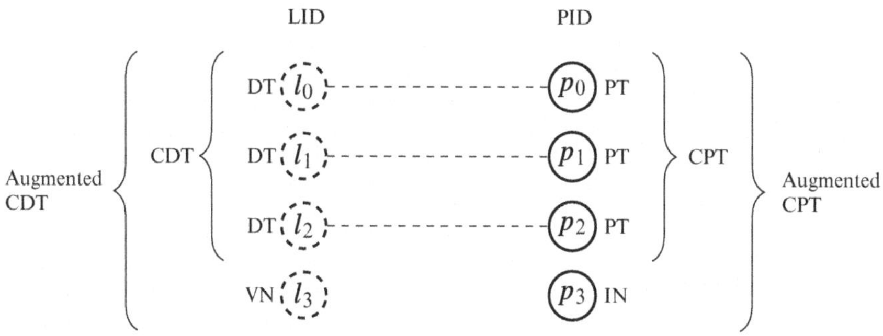

Fig. 7.6 Augmented collectives of Digital and physical twins [14]

devices (*synthesis*) and their integration in the CDT (*augmentation*). The overhead associated to the augmentation should be motivated in terms of new or improved functionality.

7.3.4 Fluidware and Digital Twins

The Fluidware vision is specifically meant to shine in highly dynamic large-scale scenarios, such as those targeted by DT solutions, as reported in Sect. 7.2.3, where adaptation to run-time contingencies and changes is the norm rather than the exception [46]. A retirement facility for elderly people or people with chronic diseases, both requiring peculiar care and continuous assistance, would be a perfect example. Once the facility is disseminated with sensor and actuator devices, and a DT layer is laid on top, one can think of exploiting the Fluidware paradigm, for example funnel processes, to deliver monitoring and control services that seamlessly cope with changes in the application: new devices (and their DTs) joining or leaving the system, new people, even new functionalities required (via composition of funnel processes).

Another appealing scenario for Fluidware, where DTs may serve the purpose of producing the event flows that Fluidware envisions, is a smart university campus. There, spaces could be reorganized on a per-need basis, such as in the case of planned temporary events (e.g. meetings, conferences, seminars, industry recruiting, etc.) or unexpected contingencies (e.g. cancellation of lectures, strikes, etc.). Services such as real-time tracking of rooms occupancy, adjustment of rooms' environmental conditions to the occupants' needs, a navigation system to steer people towards their destination featuring integrated contextual information such as current or future crowded areas are all examples of functionalities having high dynamicity and calling for adaptation, hence suitable for the Fluidware approach.

7.4 Discussion

In this section we report about challenges that despite the effort from the research community remain still open and that need to be addressed together with the presentation of some concepts that aim at enhancing the one of Digital Twin.

7.4.1 Challenges

Several challenges relate to the design and implementation of DT solutions [59]. In the following we do not report an exhaustive list, but we focus on those that in our opinion are among the most influential.

Self-Development Digital Twins are not static entities but dynamic models that evolve over time. Digital Twins are also being referred to as “living models” that continually adapt to changes in the environment or operation using real-time sensory data and can forecast the future of the corresponding physical assets [36, 43]. Therefore a way to introduce individual and collective self-development capabilities [39, 40] is required.

Dependability Digital Twin can be used for development of critical systems, with real-time constraints; therefore they require efficient interworking functions between Digital Twins and the physical systems. Different kind of DTs can be implemented in terms of resolution, complexity, modelling languages, and formats. Large-scale Digital Twin platforms are required to include distributed Digital Twin cooperation framework, flexible data-centric communication middleware, and the platform-based Digital Twin application to develop reliable systems. Many CPS vendors will provide vendor-dependent Digital Twins, and the interconnection between them is essential to make large-scale Digital Twin space [63].

Interoperability Although the literature is conceptually aligned on the idea and the importance of DT [49], there is not yet a unifying model for representing and properly working with DT across multiple application domains. The fragmentation of existing solutions is mostly related to their specificity for a target sector and creates an unnecessary substrate of heterogeneous proposals [60]. Currently, it is almost impossible to create an ecosystem where devices, services, and users can efficiently cooperate through a shared and interoperable DT vision. Nevertheless, both the industrial and academic worlds are moving towards this vision. The Industrial IoT Consortium, for instance, is proposing a shared reference architecture [44, 55] taking into account DTs relationships, composition, and main services (e.g. prediction, maintenance, safety).

The oneM2M organization[1] and the World Wide Web Consortium[2] are also actively working to provide uniform access and description of physical assets to achieve practical interoperability across multiple application domains and deployments. However, the definition and role of DTs is yet unclear in these initiatives, e.g. the WoT sees DTs as a cloud-driven interaction pattern instead of a fundamental tool to digitize and model physical assets. A broader perspective is brought by the WoDT proposal [54] that shares many aspects with the Gemini Principles vision [9]. Essentially, both aim at building DT ecosystems where many DTs can serve multiple applications' requirements in a sort of "as-a-service" paradigm.

DT Deployment The combination of IoT and DTs has traditionally happened through cloud-driven and domain-specific architectures. This limits the potential of DTs by creating an unnecessary substrate of heterogeneous implementations [60]. There, in fact, latency issues may disrupt DTs purpose, as communication between the physical and logical objects is effective only if the refresh time is lower than the average access time of applications. Thus, it should be possible to deploy DTs as close as possible to the physical devices. Fog and Edge Computing [6, 10] overcome these limitations by enabling to operate with ultra-responsive and ultra-reliable connectivity in multiple operational contexts. The adoption of distributed DTs deployed on the edge already showed some interesting results in industrial deployments [7].

However, the following limitations still hold: (i) fragmented design and modelling; (ii) adoption of monolithic and centralized approaches; (iii) limited interoperability and heterogeneity management; and (iv) Internet dependency and introduced delay. In this context, the characterization of the role, capabilities, and responsibilities of DTs running on the edge is still not well-defined, structured, and experimented in the literature. A notable exception is the work with WLDT [51], which proposes last-mile digitalization and heterogeneity management through the deployment of Edge DTs operating as active entities close to their physical counterparts.

Security Security is a challenge that needs to be addressed in all digital related sectors, but given that Digital Twins are extremely tightened with the real world, Digital Twins can be used not only to provide information for decision-making but also to make decisions and perform actions in the physical world, and security is even more relevant in reference to Digital Twins. Identifying the role that security systems and technologies such as blockchain will play is of fundamental importance for the realization and implementation of Digital Twin solutions. We can learn from studies targeting security of IoT platforms [25].

[1] https://www.onem2m.org/

[2] https://www.w3.org/

7.4.2 Advanced Digital Twin Concepts

To address some of the above challenges, new concepts that enhance the one of Digital Twin are being proposed by the literature.

Opportunistic Digital Twin The need of infrastructures focused on specific sensing tasks and powerful enough for promptly processing big amount of data hinder the development of key IoT services, especially those particularly complex and large scaled typical of smart cities. In this direction, a concept of Opportunistic Digital Twin has been defined at the confluence of DT, synthetic sensing, and Edge Intelligence (EI) aimed to simplify the (re)engineering of large-scale distributed smart systems by moving complexity from hardware infrastructure to software layer. The synergic exploitation of DT and EI techniques promises to ease both the representation and the programming of complex environments by providing straightforward metaphors for virtualizing any-scale infrastructures as well as a sound basis for injecting intelligence into its operation systems [5]. Collecting simple data from the environment and by opportunistically elaborating them through AI techniques directly at the network edge, a digital version of a physical object can be built from the bottom up as well as dynamically manipulated and operated in a data-driven manner, thus enabling prompt responses to external stimuli and effective command actuation [56].

Cognitive Digital Twin Recent advancements in IoT, big data, and machine learning have also significantly contributed to the improvements in DTs regarding their real-time capabilities and forecasting properties. Collected data constitute the so-called digital threads and are the grounding information on which simulation or machine learning algorithms rely to make predictions, enabling failures to be anticipated, to optimize the system, to design novel features, to ease and accelerate decision-making, and to improve productivity. Whenever DTs encapsulate reasoning capabilities, the concept of DT has evolved into Cognitive Digital Twin (CDT) [1, 23] that has been introduced in the literature to refer to those DTs that autonomously perform some intelligent task within the context of the physical asset, related to, e.g. smart management, maintenance, and optimization of performances. Cognitive Twin or Intelligent DTs [29] are indeed the most advanced examples of DTs and are defined just in the context of IoT Platforms and leveraging on their connectivity, analytics, simulation, and visualization tools.

7.5 Conclusion and Future Work

In this chapter we focused on Digital Twins, highlighting the interest that such a topic is attracting together with the convergence with the Fluidware project and related results. We initially reported a description of concepts related to DTs such has DT characteristics, their enabling technologies, and some usage scenarios. We then described those results of the Fluidware projects that contribute to the topic of Digital Twin. We then discussed additional challenges of DTs together with emerging concepts that enhance the concept of DTs.

From our analysis of the literature, it emerges clearly that a lot of interest is directed towards the DT topic with several research papers being published. What is missing is actually real and documented implementations of those concepts in complex scenarios. The community should start implementing what has been proposed so as to validate the approaches and highlight possible critical issues. We urge initiatives that allow different communities to work together to carry on projects in order to make the most of this innovative drive and this particular interest in the topic of Digital Twins.

References

1. Abburu, S., Berre, A.J., Jacoby, M., Roman, D., Stojanovic, L., Stojanovic, N.: COGNITWIN—hybrid and cognitive digital twins for the process industry. In: 2020 IEEE International Conference on Engineering, Technology and Innovation, ICE/ITMC 2020, Cardiff, United Kingdom, June 15–17, 2020, pp. 1–8. IEEE, Piscataway (2020).
2. Abowd, G.D.: Beyond Weiser: from ubiquitous to collective computing. Computer **49**(1), 17–23 (2016)
3. Atkinson, C., Kühne, T.: Taming the complexity of digital twins. IEEE Softw. **39**(2), 27–32 (2022)
4. Ayoobkhan, M.U.A., Yuvaraj, D., Jayanthiladevi, A., Easwaran, B., ThamaraiSelvi, R.: Smart connected digital products and IoT platform with the digital twin. In: Research Advancements in Smart Technology, Optimization, and Renewable Energy, pp. 330–350. IGI Global (2021)
5. Barbuto, V., Savaglio, C., Chen, M., Fortino, G.: Disclosing edge intelligence: a systematic meta-survey. Big Data Cogn. Comput. **7**(1), 44 (2023)
6. Bellavista, P., Berrocal, J., Corradi, A., Das, S.K., Foschini, L., Zanni, A.: A survey on fog computing for the internet of things. Pervasive Mob. Comput. **52**, 71–99 (2019)
7. Bellavista, P., Giannelli, C., Mamei, M., Mendula, M., Picone, M.: Application-driven network-aware digital twin management in industrial edge environments. IEEE Trans. Ind. Inform. **17**(11), 7791–7801 (2021)
8. Bhattacharyya, A., Izgi, E.: Digital twin technologies for high performance manufacturing. IBM White paper (2018)
9. Bolton, A., Butler, L., Dabson, I., Enzer, M., Evans, M., Fenemore, T., Harradence, F., Keaney, E., Kemp, A., Luck, A., et al.: Gemini principles (2018)
10. Bonomi, F., Milito, R.A., Zhu, J., Addepalli, S.: Fog computing and its role in the Internet of Things. In: Proceedings of the First Edition of the MCC Workshop on Mobile Cloud Computing, MCC@SIGCOMM 2012, Helsinki, Finland, August 17, 2012, pp. 13–16. ACM, New York (2012)
11. Callisto De Donato, M., Corradini, F., Fornari, F., Re, B., Romagnoli, M.: Design and development of a digital twin prototype: the SAFE case study. In: Enterprise Design, Operations, And Computing. EDOC 2023 Workshops - IDAMS, IRESEARCH, MIDas4CS, SoEA4EE, EDOC Forum, Demonstrations Track And Doctoral Consortium, Groningen, The Netherlands, October 30–November 3, 2023, Revised Selected Papers, vol. 498, pp. 107–122. Springer, Berlin (2024)
12. Callisto De Donato, M., Corradini, F., Fornari, F., Re, B., Romagnoli, M.: Enabling 3D simulation in thingsboard: a first step towards a digital twin platform. In: Enterprise Design, Operations, and Computing. EDOC 2023 Workshops - IDAMS, IRESEARCH, MIDas4CS,

SoEA4EE, EDOC Forum, Demonstrations Track And Doctoral Consortium, Groningen, The Netherlands, October 30–November 3, 2023, Revised Selected Papers, vol. 498, pp. 325–330. Springer, Berlin (2024)
13. Casadei, R.: Artificial collective intelligence engineering: a survey of concepts and perspectives. Artif. Life **29**(4), 433–467 (2023). https://doi.org/10.1162/artl_a_00408
14. Casadei, R., Pianini, D., Viroli, M., Weyns, D.: Digital twins, virtual devices, and augmentations for self-organising cyber-physical collectives. Appl. Sci. **12**(1), 349 (2021). https://doi.org/10.3390/app12010349
15. Casadei, R., Placuzzi, A., Viroli, M., Weyns, D.: Augmented collective digital twins for self-organising cyber-physical systems. In: IEEE International Conference on Autonomic Computing and Self-Organizing Systems, ACSOS 2021, Companion Volume, Washington, DC, USA, September 27–Oct. 1, 2021, pp. 160–165. IEEE, Piscataway (2021)
16. Compagnucci, I., Corradini, F., Fornari, F., Polini, A., Re, B., Tiezzi, F.: A systematic literature review on IoT-aware business process modeling views, requirements and notations. Softw. Syst. Model. **22**(3), 969–1004 (2023)
17. Corradini, F., Fedeli, A., Fornari, F., Polini, A., Re, B.: Floware: an approach for IoT support and application development. In: Enterprise, Business-Process and Information Systems Modeling—22nd International Conference, BPMDS 2021, and 26th International Conference, EMMSAD 2021, Held at CAiSE 2021, Melbourne, VIC, Australia, June 28–29, 2021, Proceedings. Lecture Notes in Business Information Processing, vol. 421, pp. 350–365. Springer, Berlin (2021)
18. Corradini, F., Fedeli, A., Fornari, F., Polini, A., Re, B.: DTMN a modelling notation for digital twins. In: Enterprise Design, Operations, and Computing. EDOC 2022 Workshops - IDAMS, SoEA4EE, TEAR, EDOC Forum, Demonstrations Track and Doctoral Consortium, Bozen-Bolzano, Italy, October 4–7, 2022, Revised Selected Papers. Lecture Notes in Business Information Processing, vol. 466, pp. 63–78. Springer, Berlin (2022)
19. Corradini, F., Fedeli, A., Fornari, F., Polini, A., Re, B.: X-IoT: a model-driven approach for cross-platform IoT applications development. In: SAC '22: The 37th ACM/SIGAPP Symposium on Applied Computing, Virtual Event, April 25–29, 2022, pp. 1448–1451. ACM, New York (2022)
20. Corradini, F., Fedeli, A., Fornari, F., Polini, A., Re, B.: Floware: a model-driven approach fostering reuse and customisation in IoT applications modelling and development. Softw. Syst. Model. **22**(1), 131–158 (2023)
21. Corradini, F., Fedeli, A., Fornari, F., Polini, A., Re, B., Ruschioni, L.: X-IoT: a model-driven approach to support IoT application portability across IoT platforms. Computing **105**(9), 1981–2005 (2023)
22. Corradini, F., Fedeli, A., Polini, A., Re, B.: Towards a digital twin modelling notation. In: IEEE International Conference on Dependable, Autonomic and Secure Computing, International Conference on Pervasive Intelligence and Computing, International Conference on Cloud and Big Data Computing, International Conference on Cyber Science and Technology Congress, DASC/PiCom/CBDCom/CyberSciTech 2022, Falerna, Italy, September 12–15, 2022. pp. 1–6. IEEE, Piscataway (2022)
23. Eirinakis, P., Kalaboukas, K., Lounis, S., Mourtos, I., Rozanec, J.M., Stojanovic, N., Zois, G.: Enhancing cognition for digital twins. In: 2020 IEEE International Conference on Engineering, Technology and Innovation, ICE/ITMC 2020, Cardiff, United Kingdom, June 15–17, 2020, pp. 1–7. IEEE, Piscataway (2020)
24. Fedeli, A., Fornari, F., Polini, A., Re, B., Torres, V., Valderas, P.: FloBP: a model-driven approach for developing and executing IoT-enhanced business processes. In: Software And Systems Modeling, pp. 1–30 (2024)
25. Fortino, G., Guerrieri, A., Pace, P., Savaglio, C., Spezzano, G.: IoT platforms and security: an analysis of the leading industrial/commercial solutions. Sensors **22**(6), 2196 (2022)
26. Fortino, G., Re, B., Viroli, M., Zambonelli, F.: Fluidware: An approach towards adaptive and scalable programming of the IoT. In: Models, Languages, and Tools for Concurrent and Distributed Programming - Essays Dedicated to Rocco De Nicola on the Occasion of His 65th Birthday. Lecture Notes in Computer Science, vol. 11665, pp. 411–427. Springer, Berlin (2019)

27. Fortino, G., Savaglio, C.: Integration of Digital Twins & Internet of Things, pp. 205–225. Springer, Berlin (2023)
28. Fortino, G., Savaglio, C.: Integration of Digital Twins & Internet of Things, pp. 205–225 (2023)
29. Grieves, M.: Intelligent digital twins and the development and management of complex systems. Digital Twin **2**(8), 8 (2022)
30. Grieves, M., Vickers, J.: Digital twin: Mitigating unpredictable, undesirable emergent behavior in complex systems. In: Transdisciplinary Perspectives on Complex Systems: New Findings and Approaches, pp. 85–113 (2017)
31. Grieves, M.W.: Digital twins: past, present, and future. In: The Digital Twin, pp. 97–121. Springer, Berlin (2023)
32. Han, J., Li, M., Guo, L.: Soft control on collective behavior of a group of autonomous agents by a shill agent. J. Syst. Sci. Complex **19**(1), 54–62 (2006)
33. He, Y., Guo, J., Zheng, X.: From surveillance to digital twin: challenges and recent advances of signal processing for industrial Internet of Things. IEEE Signal Process. **35**(5), 120–129 (2018)
34. Hoffmann, J., Heimes, P., Senel, S.: IoT platforms for the internet of production. IEEE Internet Things J. **6**(3), 4098–4105 (2019)
35. Jones, D., Snider, C., Nassehi, A., Yon, J., Hicks, B.: Characterising the digital twin: a systematic literature review. J. Manufact. Sci. Technol. **29**, 36–52 (2020)
36. Kaur, M.J., Mishra, V.P., Maheshwari, P.: The convergence of digital twin, IoT, and machine learning: transforming data into action. In: Digital Twin Technologies and Smart Cities, pp. 3–17 (2020)
37. Larsen, P.G., Fitzgerald, J., Woodcock, J.: How do we engineer trustworthy digital twins? In: Research Directions: Cyber-Physical Systems, p. 1–6 (2023)
38. Li, W., Shen, W.: Swarm behavior control of mobile multi-robots with wireless sensor networks. J. Netw. Comput. Appl. **34**(4), 1398–1407 (2011)
39. Lippi, M., Mariani, S., Martinelli, M., Zambonelli, F.: Individual and collective self-development: concepts and challenges. In: Proceedings of the 17th Conference on Computer Science and Intelligence Systems, FedCSIS 2022, Sofia, Bulgaria, September 4–7, 2022. Annals of Computer Science and Information Systems, vol. 30, pp. 15–21 (2022)
40. Lippi, M., Mariani, S., Zambonelli, F.: Developing a "sense of agency" in IoT systems: preliminary experiments in a smart home scenario. In: 19th IEEE International Conference on Pervasive Computing and Communications Workshops and other Affiliated Events, PerCom Workshops 2021, Kassel, Germany, March 22–26, 2021, pp. 44–49. IEEE, Piscataway (2021)
41. Liu, Y., Zhang, L., Yang, Y., Zhou, L., Ren, L., Wang, F., Liu, R., Pang, Z., Deen, M.J.: A novel cloud-based framework for the elderly healthcare services using digital twin. IEEE Access **7**, 49088–49101 (2019)
42. Liu, Y., Wang, Z., Han, K., Shou, Z., Tiwari, P., Hansen, J.H.L.: Sensor fusion of camera and cloud digital twin information for intelligent vehicles. In: IEEE Intelligent Vehicles Symposium, IV 2020, Las Vegas, NV, USA, October 19 - November 13, 2020, pp. 182–187. IEEE, Piscataway (2020)
43. Liu, Z., Meyendorf, N., Mrad, N.: The role of data fusion in predictive maintenance using digital twin. In: AIP Conference Proceedings, vol. 1949. AIP Publishing (2018)
44. Malakuti, S., Grüner, S.: Architectural aspects of digital twins in IIoT systems. In: Proceedings of the 12th European Conference on Software Architecture: Companion Proceedings, ECSA 2018, Madrid, Spain, September 24–28, 2018, pp. 12:1–12:2. ACM, New York (2018)
45. Mariani, S., Cabri, G., Zambonelli, F.: Coordination of autonomous vehicles: taxonomy and survey. ACM Comput. Surv. **54**(1), 19:1–19:33 (2022)
46. Mariani, S., Casadei, R., Fornari, F., Fortino, G., Pianini, D., Re, B., Russo, W., Savaglio, C., Viroli, M., Zambonelli, F.: Case studies for a new IoT programming paradigm: fluidware. In: Proceedings of the 1st Workshop on Artificial Intelligence and Internet of Things co-located with the 18th International Conference of the Italian Association for Artificial Intelligence (AI*IA 2019), Rende (CS), Italy, November 22, 2019. CEUR Workshop Proceedings, vol. 2502, pp. 82–96. CEUR-WS.org (2019)

47. Mariani, S., Picone, M., Ricci, A.: Agents and digital twins for the engineering of cyber-physical systems: opportunities, and challenges. In: Annals of Mathematics and Artificial Intelligence, pp. 1–22 (2023)
48. Minerva, R., Lee, G.M., Crespi, N.: Digital twin in the IoT context: a survey on technical features, scenarios, and architectural models. Proc. IEEE **108**(10), 1785–1824 (2020)
49. Minerva, R., Lee, G.M., Crespi, N.: Digital twin in the IoT context: a survey on technical features, scenarios, and architectural models. Proc. IEEE **108**(10), 1785–1824 (2020)
50. OMG: Business Process Model and Notation (BPMN V 2.0) (2011)
51. Picone, M., Mamei, M., Zambonelli, F.: A flexible and modular architecture for edge digital twin: implementation and evaluation. ACM Trans. Internet Things **4**(1), 1–32 (2023)
52. Picone, M., Mariani, S., Mamei, M., Zambonelli, F., Berlier, M.: WIP: preliminary evaluation of digital twins on MEC software architecture. In: 22nd IEEE International Symposium on a World of Wireless, Mobile and Multimedia Networks, WoWMoM 2021, Pisa, Italy, June 7–11, 2021, pp. 256–259. IEEE, Piscataway (2021)
53. Qi, Q., Tao, F., Hu, T., Anwer, N., Liu, A., Wei, Y., Wang, L., Nee, A.: Enabling technologies and tools for digital twin. J. Manuf. Syst. **58**, 3–21 (2021)
54. Ricci, A., Croatti, A., Mariani, S., Montagna, S., Picone, M.: Web of digital twins. ACM Trans. Internet Tech. **22**(4), 101:1–101:30 (2022)
55. da S. Souza, V., da Cruz, R.G., Silva, W., Lins, S., Lucena, V.: A digital twin architecture based on the industrial internet of things technologies. In: IEEE International Conference on Consumer Electronics, ICCE 2019, Las Vegas, NV, USA, January 11–13, 2019, pp. 1–2. IEEE, Piscataway (2019)
56. Savaglio, C., Barbuto, V., Awan, F.M., Minerva, R., Crespi, N., Fortino, G.: Opportunistic digital twin: an edge intelligence enabler for smart city. ACM Transactions on Sensor Networks (2023)
57. Semeraro, C., Lezoche, M., Panetto, H., Dassisti, M.: Digital twin paradigm: a systematic literature review. Comput. Ind. **130**, 103469 (2021)
58. Shahat, E., Hyun, C.T., Yeom, C.: City digital twin potentials: a review and research agenda. Sustainability **13**(6) (2021)
59. Sharma, A., Kosasih, E.E., Zhang, J., Brintrup, A., Calinescu, A.: Digital twins: State of the art theory and practice, challenges, and open research questions. J. Ind. Inf. Integr. **30**, 100383 (2022). https://doi.org/10.1016/j.jii.2022.100383
60. Tao, F., Qi, Q.: Make more digital twins. Nature **573**, 490–491 (2019)
61. Tuong, N., Jump, A., Casey, D.: Emerging Tech Impact Radar: 2023: Gartner Research Excerpt. https://www.gartner.com/en/doc/emerging-technologies-and-trends-impact-radar-excerpt
62. Uhlemann, T.H.J., Lehmann, C., Steinhilper, R.: The digital twin: realizing the cyber-physical production system for industry 4.0. Proc. CIRP **61**, 335–340 (2017), the 24th CIRP Conference on Life Cycle Engineering
63. Yun, S., Park, J.H., Kim, W.T.: Data-centric middleware based digital twin platform for dependable cyber-physical systems. In: 2017 Ninth International Conference on Ubiquitous and Future Networks (ICUFN), pp. 922–926. IEEE, Piscataway (2017)
64. Zambonelli, F., Viroli, M., Fortino, G., Re, B.: Towards adaptive flow programming for the IoT: The fluidware approach. In: IEEE International Conference on Pervasive Computing and Communications Workshops, PerCom Workshops 2019, Kyoto, Japan, March 11–15, 2019, pp. 549–554. IEEE, Piscataway (2019)

Chapter 8
Smart Collective Services in the Edge–Cloud Continuum Based on a Simulation-Driven Toolchain

Danilo Pianini and Claudio Savaglio

8.1 Introduction

With the advent of the Internet of Things (IoT) and the rise of edge computing, the infrastructure hosting smart services has been changing at an unprecedented pace. If once the only way to provide a service was to deploy it on a server, today the same service can be deployed on a plethora of devices, from the cloud to the edge, to the very devices that are part of the service itself. We are indeed witnessing the rise of a new paradigm, the *edge–cloud continuum* [1, 2], where services are deployed on a continuum of devices spanning across multiple networks (also referred as *Multiple IoTs* [3]).

The Edge-Cloud Continuum (ECC) [4] opens a much larger search space for the deployment of smart services; however, exploiting the new options is not trivial, as it requires the capability to investigate how a deployment would behave on a given infrastructure *in advance*, before deploying it.

As introduced in Chap. 3, the Fluidware architecture features a seamless 3-tiered organization promoting the continuum of data and services that covers IoT devices, Edge and Cloud Servers. Thus, by default, Fluidware promotes openness and flexibility regarding the target infrastructure. Still, to leverage these engineering opportunities, the designer relying on the abstractions provided by Fluidware must make sure that deployment-level concerns do not boil up to the design level. In other words, the designer must be able to specify the entire system from an abstract point

D. Pianini (✉)
Alma Mater Stodiurum—Università di Bologna, Cesena, Italy
e-mail: danilo.pianini@unibo.it

C. Savaglio
Università della Calabria, Rende (CS), Italy
e-mail: csavaglio@dimes.unical.it

F. Zambonelli et al. (eds.), *Fluidware*, Internet of Things,
https://doi.org/10.1007/978-3-031-62146-8_8

of view, and then be able, with no changes to the design *and* the implementation, to deploy it on an infrastructure that is unknown in advance.

Specifying the system behavior without considering deployment details requires methods and tools specifically designed to support this operation. In this chapter, we present a methodology and a toolchain to support the design of smart services with the Fluidware metaphor and their deployment on the ECC in a deployment-agnostic fashion. At the core of the approach is a divide-and-conquer strategy, where the logical devices that have been programmed following Fluidware are decomposed into deployable communicating units, which can then be deployed in arbitrary nodes of the ECC (as far as they meet the minimal requirements of the deployment unit). This way, multiple deployment options can be investigated prior to deployment through simulation, which becomes a key tool in the design process, rising to the role that debuggers and profilers have in traditional software development: Support the designer in understanding the behavior of the system, hunt for defects, and measure the performance.

The remainder of this chapter is as follows: Sect. 8.2 briefly discusses trends and approaches featuring the development methodologies for pervasive systems, with a focus on simulation toolchains for the IoT domain. Section 8.3 presents the methodology and specifically how Fluidware funnels can be decomposed through the pulverization approach. The main elements of the toolchain are described in Sect. 8.4, while final remarks conclude the chapter in Sect. 8.5.

8.2 State of the Art

Pervasive systems are characterized by their ability to seamlessly integrate technology into everyday life as well as to operate in diverse and dynamic environments, often involving a wide range of devices (sensors, actuators, cameras, just to name a few), networks, and software (adapters, middleware, proxy, etc.). Hence, developing IoT systems provided with such a degree of pervasiveness requires careful consideration of various factors (e.g., context awareness, adaptability, scalability, and user experience) including strategies to figure out the impact on the deployment on the performance of the system to be [5]. Therefore, over the years, an increasing number of methodologies arise, aimed at providing guidelines, best practices, workflows to support the complex and error-prone development process. Mostly pivoted around well-established approaches like Model-Driven Development (MDD), Agile development, User-Centered Design, and Service-Oriented Architecture (SOA) principles [6], these methodologies present both qualitative and quantitative solutions, often multidisciplinary, highly prescriptive, and extensively documented [7]. The constitutive activities and procedures provide consistency by specifying "what" steps to take, "when" and "how" to perform those steps but, most importantly, the reasons "why" those steps should be taken, in particular order: In such a way, methodologies provide philosophical (i.e., components' gluing and development approach) foundations for IoT systems but also a technical guidance (i.e., which paradigm and tools to adopt), since the deployment configuration and differences among available implementations are too fundamental to be ignored. As a matter of fact, evaluating the performance of a collective adaptive application prior to its actual deployment is key but not easy due to the multiplicity, heterogeneity, and context-dependent metrics to be considered, reason why most classical testing procedures are not readily applicable.

If, on one hand, methodologies need to be full-fledged and comprehensive to consider design, implementation, and deployment aspects for a truly reliable support at the development process, on the other hand, it would be highly desiderata to separate development and deployment, so to leave the developer free to focus on the operational aspects of the system, disregarding deployment details and exploring several configuration options through simulation. Indeed, simulation can be used in a robust way to such purpose, since solutions can be selected not only based on their outcome with the expected value of the parameters, but also taking into account the reliability of the behavior in a wide range of situations. Therefore, simulation plays an important role in the overall system development process, and it is expected to be embedded within toolchain of the development methodologies [8]. Recently, simulators designed specifically for comparative analysis between edge- and cloud-based deployment of an IoT system have been gaining attention and evaluated in terms of both metrics (functional suitability, performance efficiency, compatibility, reliability, maintainability, portability, etc.) and modeling characteristics (modeling of infrastructure and application, resource management, mobility, and scalability). But few of them (iFEdgeSim, EdgeCloudSim, EdgeSimPy, IoTSimEdge, IoTSim-Osmosis) allow simulating IoT services truly spanning across the ECC.

8.3 Methodology

The idea behind the methodology we are proposing is to separate the definition of the software system and the definition of the infrastructure where it will be deployed as neatly as possible, constrain the possible deployments, leverage simulation to obtain predeployment performance indications, and finally deploy the actual system on the target infrastructure.

8.3.1 High-Level View

In principle, Fluidware systems can be defined independently of their target infrastructure. In fact, Fluidware abstractions define communication among *logical* entities. Although typically logical entities are mapped to physical devices, this is not a requirement. Thus, the system can be defined without considering the target infrastructure and written as the designer sees fit.

Of course, however, at some point the system will need to be deployed on a physical infrastructure. Hence, there must be a way to describe it in an actionable fashion, with explicit model of the capabilities of devices that will be part of the infrastructure.

Once both the infrastructural and the behavioral descriptions are ready, automated machinery can be used to generate possible/compatible deployments. In a common Fluidware system counting hundreds or thousands of devices, these deployments will be many: How to select the "best" one? And, on a related note, how to define what "best" means? Key performance indicators can vary wildly with the application at hand: An agricultural monitoring system deployed on a large area may require a large data rate but be very tolerant in terms of latency (a few seconds may be acceptable), while a system for real-time monitoring of health parameters may have opposite requirements. Although Fluidware systems have been demonstrated to be able to tackle scenarios where requirements of convergence in time and distribution of the available events vary wildly [9], all the considerations abstract away from the actual infrastructure and the deployment performance. Thus, it is paramount that performance metrics can be expressed unambiguously and measured before deployment. In the proposed approach, the efficacy of potential deployments is evaluated via simulation, comparing the generated combinations of infrastructure and software system.

8.3.2 Decomposing Funnels via Pulverization

For the approach to be applicable, logical devices must be decomposable into small deployment units. Fortunately, the flexibility induced by the Fluidware abstractions naturally allows for decomposition under many implementations.

A possible implementation of funnels, in fact, can be designed, relying on the abstractions provided by aggregate computing [10], in terms of *aggregate processes* [11]. It has been shown that arbitrary aggregate computations can be decomposed by *pulverization* [12], an approach already introduced in Chap. 3 and briefly resumed here for reasons of self-containment. In pulverizable applications, logical devices can be *automatically* decomposed into small deployment units that can be deployed on arbitrary network nodes, in any part of the ECC. As far as the network where the logical device components are executing is not segmented, from the point of view of the designer, there will always be the network they programmed for, even if the underlying infrastructure is entirely different. These abstractions enable several interesting scenarios. For instance, the designer may program devices with limited computational capabilities (such as LoRaWAN Motes [13]) as if they were capable of executing complex computations, relying on the fact that the resource-hungry part of the application will be transparently deployed on a diverse machine. Moreover, should the technology of these devices leap forward and make them capable enough to run the complex software that was previously destined to a supporting machine, the deployment could be redefined without any change to the application logic. This also means that the same software can be ported across different technological setups: From the point of view of the application logic, there is no difference if the sensors for a certain application are deployed on a battery-run LoRa mote or on a much more capable single-board computer connected to the power grid.

A second interesting scenario is the possibility to define arbitrary logical network topologies. The designer, in fact, can reason as if devices were connected in a certain way, for instance, based on physical proximity (as they would be in an opportunistic mesh network) or based on application-level information (such as subscribed to the same group on a social network). Additionally, besides these interesting possibilities for the application designer, the approach has a very practical application and immediate application: The ability to flexibly shift the application load across the whole ECC allows for fine-grained, opportunistic, and adaptive load balancing, with custom trade-offs (e.g., moving computation on the cloud when the battery is low, as far as the cost for spawning more resources remains under check).

8.3.3 Infrastructure Models

To perform the mapping between the (pulverized) deployment units and their actual hosts automatically, bridging the gap between the logical and the physical world, a formal, machine-understandable definition of the target infrastructure is required. A natural way to express such a definition is to rely on a syntax similar to the one adopted by Infrastructure as Code (IaC) tools such as Terraform [14], with two major differences:

- The definition must capture which hardware capabilities are available, e.g., in terms of sensors and actuators.
- The configuration must capture *multiple* infrastructures at once (those among which the actual system may be deployed).

Regarding the first point, some deployment units may require specific hardware capabilities to be available on the host. For instance, a deployment unit that provides a video stream may require a camera to be available, and a deployment unit that performs intensive computations may require a sufficiently powerful CPU: If the infrastructure includes a smartphone with a camera and the cloud, we can expect the first deployment unit to be bound to run on the smartphone and the latter to be able to be deployed either in the phone or in the cloud.

Concerning the expression of multiple infrastructures at once, a reasonable starting point here is a collection of IaC configuration files, each expressing a viable configuration. However, with multiple options at hand, the complexity of the approach (and the file count) grows combinatorially; thus, as soon as the system must select between several deployments, a dedicated language should be concocted, capable to express the whole space of available infrastructures.

8.3.4 Deployment Generation

Once the application has been decomposed via pulverization and a valid description of the target infrastructure(s) is available, the system can automatically generate the assignment between the pulverized component (or deployment unit) and actual hosts. In this phase, we are looking at a plethora of possible valid mappings between hosts and deployment units.

In complex cases, it would be reasonable that they do not get generated ahead of time, but progressively as the exploration of this solution space progresses. Progressive deployment generation can be optimized by leveraging an evaluation system with appropriate metrics: This way, the search for good deployments becomes a matter of running a classic optimization procedure, where the computation of one point in the search space implies the execution of a simulated deployment. In turn, these direct the generation of new deployments, skipping the less promising ones.

8.4 Toolchain

Once Fluidware applications are expressed using a pulverizable approach (such as aggregate processes), we need additional tools to formally describe and exercise the deployments in a controlled environment before moving to production systems. In this section, we introduce possible deployment description tools and viable simulation systems.

8.4.1 Deployment Description Tools

Formally describing a deployment requires two macro-steps: describing the target infrastructure and describing the component allocation.

The definition of a target infrastructure is a matter of defining the available hardware resources. Although built for a different purpose, Infrastructure as Code (IaC) tools provide guidance to this end. Ideally, a target infrastructure deployment tool could extend the syntax of tools such as Terraform [14], providing means to define more than a static configuration of resources, and rather a set of possible configurations.

Such tool could be complemented with a dedicated tool to express the component allocation for every deployment, or, possibly, the two could be merged into a single tool. So far, in existing experiments, the infrastructure and allocation languages have been kept separated [12, 15]. However, promising approaches such as Multitier Programming may provide ways to merge the two [16].

8.4.2 Integration on Existing Simulation Systems

Assessing the performance of a collective adaptive application before deploying it is crucial but challenging. This is because such applications involve multiple and potentially diverse entities, making traditional testing methods less straightforward. Furthermore, these systems are typically evaluated using various continuous metrics across different scenarios, often relying on experiential and estimated data. In such cases, simulation becomes a robust solution, allowing decisions to be based not only on expected outcomes but also on the reliability of behavior in various situations. As a result, simulation plays a significant role in the overall system engineering process, especially in verifying if prototypes meet functional requirements. Simulators designed for prototyping and development tend to abstract away network-level and nonapplication-level details, prioritizing scalability and system behavior debugging. Commonly used simulators during the software design phase include Repast [17], NetLogo [18], and Alchemist [19], with the latter providing strong support for aggregate computing specifications defined in Protelis [20] or ScaFi [21].

Once a system has been functionally verified, the assessment needs to shift toward low-level aspects like networking and power consumption. It is important to ensure that the chosen solution can effectively support the system in real-world deployment, as it may encounter issues not apparent in abstract simulated models. In such cases, there are two options: either exploring alternative implementation choices at the application level or finding more efficient ways to map the application logic onto the deployed system. Systems that support pulverization provide a systematic approach to address these challenges, separating functional logic from deployment specifics and enabling the exploration of various configuration options.

Exploring these configurations necessitates specialized tools that reveal deployment details and accurately capture resource consumption information. These tools have traditionally included network simulators like OMNeT++ [22] and NS-3 [23], operating-system-level simulators such as TOSSIM [24], and cloud-oriented simulators like CloudSim [25], DEVS [26], and Recap [27]. Recently, there has been a growing interest in simulators designed specifically for comparative analysis between edge- and cloud-based deployments of IoT systems. These technologies have been reviewed in [28, 29] and categorized based on ISO/IEC 25010/25023 Quality/Metrics Characteristics (e.g., functional suitability, performance efficiency, compatibility, reliability, maintainability, portability) and modeling features (including infrastructure, application, and resource management modeling, scalability, and mobility models). From both surveys, EdgeCloudSim [30] emerges as one of the richest edge–cloud simulators. EdgeCloudSim is a simulation toolkit that extends CloudSim to cater to the unique characteristics of ECC. It allows users to model and simulate devices, edge nodes, and cloud data centers, enabling the evaluation of various resource management policies and application performance, also in the presence of different networking models featured by their own latency, bandwidth, and reliability characteristics. As additional adaptivity, EdgeCloudSim is designed to be scalable and flexible, enabling the simulation of large-scale ECC scenarios, allowing researchers to experiment with various resource management policies, such as task offloading strategies, load balancing algorithms, and energy-aware optimizations.

8.5 Conclusion

Developing complex and dynamic IoT systems in uncertain environments is a significant engineering challenge. These systems require programming paradigms that support autonomy, decentralization, and adaptiveness from a functional perspective. On the deployment side, due to various factors related to computation, networking, and mobility, they can be deployed in various ways, some of which might not be optimal, leading to ineffective configurations and reconfigurations. To address this, computing paradigms that separate logic and deployment, along with simulation tools for evaluating different deployment configurations, play a vital role in the efficient development of complex IoT systems.

In this chapter, we have presented a methodology and toolchain for capturing potential deployments of intelligent edge services and predicting their performance and cost through simulation. This solution is tailored for approaches that allow for logical partitioning and deployment across the IoT–Edge–Fog–Cloud infrastructure continuum, and, in particular, they assume Fluidware applications with funnels implemented as aggregate processes. Based on the same logical specification, different deployment configurations can yield varying system performance, depending on factors like users' number and resource availability. This methodology can provide advance estimates of system performance, enabling operational teams to choose the most suitable deployment platform or integrate it into automated deployment processes for quality control.

References

1. Milojicic, D.: The edge-to-cloud continuum. Computer **53**, 16–25 (2020)
2. Moreschini, S., Pecorelli, F., Li, X., Naz, S., Hästbacka, D., Taibi, D.: Cloud continuum: the definition. IEEE Access **10**, 131876–131886 (2022)
3. Cauteruccio, F., Cinelli, L., Corradini, E., Terracina, G., Ursino, D., Virgili, L., Savaglio, C., Liotta, A., Fortino, G.: A framework for anomaly detection and classification in multiple IoT scenarios. Fut. Gener. Comput. Syst. **114**, 322–335 (2021)
4. Bittencourt, L., Immich, R., Sakellariou, R., Fonseca, N., Madeira, E., Curado, M., Villas, L., DaSilva, L., Lee, C. & Rana, O. The Internet of Things, fog and cloud continuum: integration and challenges. Internet Things **3**, 134–155 (2018)
5. Becker, C., Julien, C., Lalanda, P., Zambonelli, F.: Pervasive computing middleware: current trends and emerging challenges. CCF Trans. Pervasive Comput. Interaction **1**, 10–23 (2019)
6. Sosa-Reyna, C., Tello-Leal, E., Lara-Alabazares, D.: Methodology for the model-driven development of service oriented IoT applications. J. Syst. Archit. **90**, 15–22 (2018)

7. Fortino, G., Savaglio, C., Spezzano, G., Zhou, M. Internet of Things as system of systems: a review of methodologies, frameworks, platforms, and tools. IEEE Trans. Syst. Man Cybern. Syst. **51**, 223–236 (2021)
8. Savaglio, C., Fortino, G.: A simulation-driven methodology for IoT data mining based on edge computing. ACM Trans. Internet Technol. **21**, 1–22 (2021)
9. Pianini, D., Casadei, R., Viroli, M., Mariani, S., Zambonelli, F.: Time-fluid field-based coordination through programmable distributed schedulers. Log. Methods Comput. Sci. **17** (2021). https://doi.org/10.46298/lmcs-17(4:13)2021
10. Beal, J., Pianini, D., Viroli, M.: Aggregate programming for the Internet of Things. Computer **48**, 22–30 (2015). https://doi.org/10.1109/MC.2015.261
11. Casadei, R., Viroli, M., Audrito, G., Pianini, D., Damiani, F.: Engineering collective intelligence at the edge with aggregate processes. Eng. Appl. Artif. Intell. **97**, 104081 (2021). https://doi.org/10.1016/j.engappai.2020.104081
12. Casadei, R., Pianini, D., Placuzzi, A., Viroli, M., Weyns, D.: Pulverization in cyber-physical systems: engineering the self-organizing logic separated from deployment. Future Internet **12**, 203 (2020). https://doi.org/10.3390/fi12110203
13. Sherazi, H., Grieco, L., Imran, M., Boggia, G.: Energy-efficient LoRaWAN for industry 4.0 applications. IEEE Trans. Ind. Inform. **17**, 891–902 (2021). https://doi.org/10.1109/TII.2020.2984549
14. Howard, M.: Terraform—automating infrastructure as a service (2022). CoRR. abs/2205.10676. https://doi.org/10.48550/arXiv.2205.10676
15. Casadei, R., Fortino, G., Pianini, D., Placuzzi, A., Savaglio, C., Viroli, M.: A methodology and simulation-based toolchain for estimating deployment performance of smart collective services at the edge. IEEE Internet Things J. **9**, 20136–20148 (2022). https://doi.org/10.1109/JIOT.2022.3172470
16. Aguzzi, G., Casadei, R., Pianini, D., Salvaneschi, G., Viroli, M.: Towards pulverised architectures for collective adaptive systems through multi-tier programming. In: IEEE International Conference on Autonomic Computing and Self-Organizing Systems, ACSOS 2021, Companion Volume, Washington, DC, USA, September 27–Oct. 1, 2021, pp. 99–104 (2021). https://doi.org/10.1109/ACSOS-C52956.2021.00033
17. North, M., Collier, N., Vos, J.: Experiences creating three implementations of the repast agent modeling toolkit. ACM Trans. Model. Comput. Simul. **16**, 1–25 (2006). https://doi.org/10.1145/1122012.1122013
18. Sklar, E.: NetLogo, a multi-agent simulation environment. Artif. Life. **13**, 303–311 (2007). https://doi.org/10.1162/artl.2007.13.3.303
19. Pianini, D., Montagna, S., Viroli, M.: Chemical-oriented simulation of computational systems with ALCHEMIST. J. Simul. **7**, 202–215 (2013). https://doi.org/10.1057/jos.2012.27
20. Pianini, D., Viroli, M., Beal, J.: Protelis: practical aggregate programming. In: Proceedings of the 30th Annual ACM Symposium on Applied Computing, Salamanca, Spain, April 13–17, 2015, pp. 1846–1853 (2015). https://doi.org/10.1145/2695664.2695913
21. Casadei, R., Viroli, M., Aguzzi, G., Pianini, D.: ScaFi: a scala DSL and toolkit for aggregate programming. SoftwareX **20**, 101248 (2022). https://doi.org/10.1016/j.softx.2022.101248
22. Varga, A., Hornig, R.: An overview of the OMNeT++ simulation environment. In: Proceedings of the 1st International Conference on Simulation Tools and Techniques for Communications, Networks and Systems & Workshops, SimuTools 2008, Marseille, France, March 3–7, 2008, p. 60 (2008). https://doi.org/10.4108/ICST.SIMUTOOLS2008.3027
23. Riley, G., Henderson, T.: The <i>ns-3</i> network simulator. In: Modeling and Tools for Network Simulation, pp. 15–34 (2010). https://doi.org/10.1007/978-3-642-12331-3%5C_2
24. Levis, P., Lee, N., Welsh, M., Culler, D.: TOSSIM: accurate and scalable simulation of entire tinyOS applications. In: Proceedings of the 1st International Conference on Embedded Networked Sensor Systems, SenSys 2003, Los Angeles, California, USA, November 5–7, 2003, pp. 126–137 (2003). https://doi.org/10.1145/958491.958506
25. Calheiros, R., Ranjan, R., Beloglazov, A., Rose, C., Buyya, R.: CloudSim: a toolkit for modeling and simulation of cloud computing environments and evaluation of resource provisioning algorithms. Softw. Pract. Exp. **41**, 23–50 (2011). https://doi.org/10.1002/spe.995

26. Chow, A., Zeigler, B.: Parallel DEVS: a parallel, hierarchical, modular, modeling formalism. In: Proceedings of the 26th Conference on Winter Simulation, WSC 1994, Lake Buena Vista, FL, USA, December 11–14, 1994, pp. 716–722 (1994). https://doi.org/10.1109/WSC.1994.717419
27. Byrne, J., Svorobej, S., Gourinovitch, A., Elango, D., Liston, P., Byrne, P., Lynn, T.: RECAP simulator: simulation of cloud/edge/fog computing scenarios. In: 2017 Winter Simulation Conference, WSC 2017, Las Vegas, NV, USA, December 3–6, 2017, pp. 4568–4569 (2017). https://doi.org/10.1109/WSC.2017.8248208
28. Svorobej, S., Endo, P., Bendechache, M., Filelis-Papadopoulos, C., Giannoutakis, K., Gravvanis, G., Tzovaras, D., Byrne, J., Lynn, T.: Simulating fog and edge computing scenarios: an overview and research challenges. Future Internet **11**, 55 (2019). https://doi.org/10.3390/fi11030055
29. Ashouri, M., Lorig, F., Davidsson, P., Spalazzese, R.: Edge computing simulators for IoT system design: an analysis of qualities and metrics. Future Internet **11**, 235 (2019). https://doi.org/10.3390/fi11110235
30. Casadei, R., Placuzzi, A., Viroli, M., Weyns, D.: Augmented collective Digital Twins for self-organising cyber-physical systems. In: IEEE International Conference on Autonomic Computing and Self-Organizing Systems, ACSOS 2021, Companion Volume, Washington, DC, USA, September 27–Oct. 1, 2021, pp. 160–165 (2021)

Chapter 9
Envisioning Unpredictability in Smart Environments

Danilo Pianini, Barbara Re, Lorenzo Rossi, and Franco Zambonelli

9.1 Introduction

IoT systems, composed of devices sensing and acting on the environment, are increasingly adopted in several sectors (e.g., agriculture and manufacturing). IoT systems can be exploited to achieve a specific environment state, a collaborative goal, by coordinating their components in a process-oriented manner [12]. However, the dynamicity of the real world creates unpredictable situations, such as sensor failures, actuation errors, energy consumption constraints, and the addition of new devices at runtime. Such unpredictability requires the design and development of IoT systems to be capable of recognizing situations and adapting to contingencies. Inspired by these scenarios, we identify three main challenges to be faced, and the corresponding areas of work that we believe should be tackled jointly, as they are intertwined.

The first challenge consists of infusing the system with a **causal sense of agency**, namely its ability to recognize the effects of its own actions on the environment and interactions among devices. Thus, it is necessary to enable self-perception, focusing on unpredictable changes and evolution. This can also provide insights into the effect of actions and reconfigurations on the evolution of the overall system.

D. Pianini
Alma Mater Stodiurum—Università di Bologna, Cesena, Italy
e-mail: danilo.pianini@unibo.it

B. Re · L. Rossi (✉)
University of Camerino, Camerino, Italy
e-mail: barbara.re@unicam.it; lorenzo.rossi@unicam.it

F. Zambonelli
University of Modena e Reggio Emilia, Reggio Emilia, Italy
e-mail: franco.zambonelli@unimore.it

F. Zambonelli et al. (eds.), *Fluidware*, Internet of Things,
https://doi.org/10.1007/978-3-031-62146-8_9

The second challenge refers to moving from the statically designed process to hierarchical collaboration processes that **(self-)compose** in a meaningful way. This should be made possible by advancing the state of the art by prompting the self-integration of unknown information sources (i.e., novel devices) and the hierarchical composition of existing components into larger ones. This enables system designers to keep raising complexity under control, de facto introducing inspectable hierarchical abstraction levels.

The last challenge consists of taking a business view where IoT devices should be at the service of **collaborative goals** and appropriately integrated in a process-oriented manner. This challenge can be faced by exploiting a distributed coordination mechanism closing the gap between the top-down, process-driven view and the bottom-up, self-organizing view. This can lead to harmonizing global and strategic views (where the business goals live) with a bottom-up exploration of devices composing the smart environments.

To fulfill this view, in this chapter we envision Bidirectional Responsive IoT Collaboration and Knowledge System (BRICKS), a novel computational model for adaptable and open IoT systems. BRICKS recognizes the system's contingencies and autonomously reconfigures its behavioral processes to steer the entire system toward globally specified goals.

The remainder of this chapter is organized as follows: Sect. 9.2 defines a reference application context, Sect. 9.3 discusses challenges and sketches the BRICKS approach and how it would address those challenges, Sect. 9.4 defines the path toward the full development of BRICKS and the research activities required, Sect. 9.5 discusses related work in the area, and finally Sect. 9.6 summarizes the contributions of this chapter and discusses the ongoing developments.

9.2 Motivating Scenario

The BRICKS approach aims to be portable to any scenario where situation recognition and dynamic response to unpredicted events are paramount. These are becoming typical of several sectors as they progressively digitalize: Healthcare, public transportation, smart agriculture, and smart manufacturing are activities that are increasingly adopting smart environments and are thus likely targets for the proposed techniques.

Among them, to provide an example and for the sake of clarity, we will focus on the context of food production, ranging from precision agriculture to the food conservation and delivery chain. In precision farming, novel techniques are used to control the activities at the intrafield level. Fine-grained monitoring of the environmental conditions (temperature, humidity, soil acidity, pest activity, etc.), combined with precision actuation (local irrigation and fertilization, greenhouse ventilation, per-plant pest control, etc.), can promote resource-saving and, at the same time, increase yield. Of course, food production does not conclude with harvesting: The supply chain includes food processing, storage, and delivery. All these related phases (including farming) must abide by strict safety and sanitary regulations (regarding, for instance, the maximum allowed use of pesticides, the storage temperature, the transport times, and the hygienic conditions). Precision agriculture techniques are pushing the boundaries of complexity: IoT technologies with robotic actuators may help directly in pest control, quality assurance, and crop collection, but there is also a much undervalued secondary impact; the unprecedented amount of data obtained by a precisely monitored field can potentially impact the organization of the whole supply chain. Of course, this kind of cascading propagation of knowledge is sensitive to changes in the available information. In particular, it is subject to multiple sources of dynamicity (changes in the smart environment configuration, new data, new sensors, malfunctions, complexities related to the logistics, changes in the regulations, etc.): As a result, its efficacy is heavily influenced by the ability of systems to cope with a dynamic environment, face unforeseen events, and pursue a shared objective.

We now introduce a running example that will accompany us along the presentation of the BRICKS approach and its compositional model. The example comes from a precision agriculture scenario in which a robotic arm and several sentinel devices collaborate to manage a greenhouse. Each sentinel device monitors the conditions of a plant inside the greenhouse. Every 30 minutes, the sentinels read their sensors to gather information about the plant they are responsible for. This information includes the plant's position and other data useful for determining which treatments to apply.

The information about a plant is sent to the robotic arm that, upon the reception of new signals, decides which treatments to implement on a plant, generating a job. As long as the list of jobs is not empty, the robotic arm executes them one by one.

Specifically, for each job, the robotic arm reaches the position of the plant to be treated, performs the necessary treatments, and then moves on to the next job. The activity ends when there are no more jobs on the list and resumes when new jobs arise. Figure 9.1 shows a representative image of the case study.

9.3 BPMN Meets the IoT: the BRICKS Approach

The BRICKS approach, graphically summarized in Fig. 9.2, envisions IoT systems where *causal sense of agency*, *self-composition*, and *collaboration to reach a collective goal* coexist and reinforce each other, leading to augmented processes [19]. The system designer can work to define the system behavior top-down, based on process models. At the same time, whenever new possibilities arise (for instance, new sensors or actuators get connected), the system autonomously reacts and produces new composable behavioral blocks rooted in the perception of the causal sense of agency. These composable system blocks, which we call Bricks (Bricks), can self-integrate hierarchically into Bricks of Bricks that become available to modify or enhance the existing IoT system.

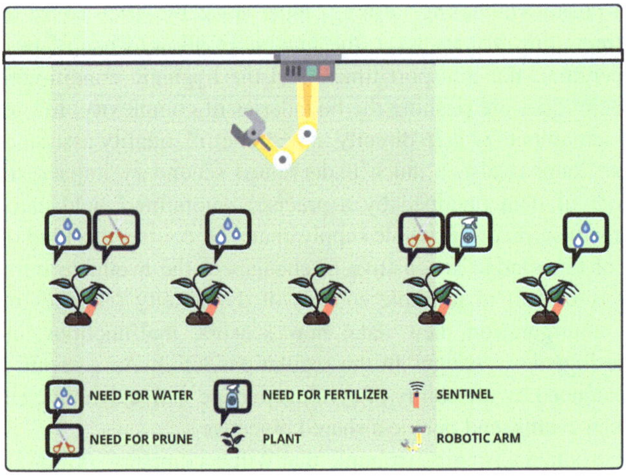

Fig. 9.1 Representation of the running example

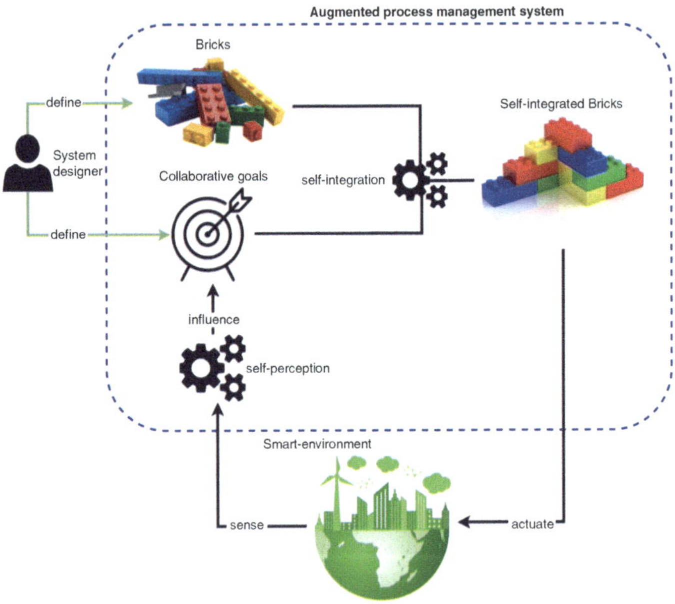

Fig. 9.2 Overview of the BRICKS approach

Crucially, novel composition approaches for such dynamic IoT systems must be scalable (able to deal with many devices) and robust, as the inherent openness may cause devices to join or leave (permanently or temporarily).

Accounting for new devices is particularly challenging, as they may expose the IoT systems to a new and unknown interface, which introduces complexity on two levels: first, keeping the preexisting functionality nominal requires the ability to adapt to the new way of accessing resources; second, the new interface may provide an entirely new set of possibilities, which in turn may trigger a system-wide adaptation process that goes beyond the compliance to the new interface and opens new possibilities for achieving collaborative goals. Openness is inextricably entwined with unpredictability, in both context and situation recognition and data production and processing. From a practical point of view, data structures can hardly be predefined by designing for every possible future event. As such, data is generally considered unstructured, making it impractical, when not entirely impossible, to provide a predefined mapping between observations and (re)actions that can be performed on the environment.

An additional challenge of primary relevance is tackling heterogeneity. The BRICKS approach's systems of interest are inherently heterogeneous on multiple levels: different devices, different services (yet possibly responding to similar needs), different protocols, and different applications. When hierarchical composition enters the arena, heterogeneity must be brought up to the collaborative goal level. This, in turn, heralds potential conflicts among system components. Consequently, the composition must be carefully crafted considering the context in which the objective for the composed object is being constructed. Scalability, robustness, and heterogeneity, traditional characteristics of an open and dynamic environment, can benefit from the approach results.

The innovative idea of the BRICKS approach is that IoT systems can be realized through self-integrating Bricks, which, in addition to the role of data prosumers, can integrate with other Bricks, generating new Bricks that represent their composition. The driving forces behind Bricks composition can be exogenous when a system designer defines some Bricks, or endogenous (which we also refer to with the name of self-integration) when combinations of Bricks derive from an observation of the smart environment.

In both cases, an evaluation of the new Bricks structure is necessary before it can be safely introduced in the runtime. Evaluation can be performed considering prespecified collaborative goals in a sandbox environment. In such a way, BRICKS will be able to establish, in a safe manner, which behavior should be reflected in the real environment.

Technically, we see Bricks as the merger of other concepts existing in computer science, enriched with a novel part. Their behavior as prosumers and their ability to learn resemble agents, but they are not necessarily proactive, and they require different composability capabilities. Isolation, the ability to "wrap" existing software, and agnosticism to the deployment target are inspired by containers; however, with respect to them, their composition necessarily creates a new Brick, enabling hierarchical structures. The hierarchical composition is similar to both the hierarchical modeling of business process notation (enabled by the concept of subprocess) and the holonic control architectures [18], but Bricks are enriched with learning and self-agency, and compositions can arise automatically. Finally, the combination of distributed Bricks largely resembles the functional "building blocks" composition proposed in frameworks such as aggregate programming [2].

However, entities participating in the composition process fundamentally differ from aggregate algorithms as they are evolving entities, and the composition itself can be spontaneous.

9.4 Toward the Realization of the BRICKS Approach

In this section, we discuss the main ingredients, and the recipe that we expect can lead to the realization of the BRICKS approach. The specification of single Bricks and Bricks of Bricks will be discussed considering three dimensions:

1. Structural (how the Brick is made)
2. Behavioral (what the Brick does), and
3. Interactional (how multiple Bricks interact)

9.4.1 Compositional Model

A Brick is a computational unit that refers to a basic activity of an IoT device. This involves the elicitation of the device capabilities under four categories:

1. Sense
2. Actuate/Move
3. Communicate, and
4. Compute

Sense actions involve the perception of the surroundings, e.g., capture an image with a camera, or a sound with a microphone, or read the current speed of an electric engine. Actuate/Move includes actions of devices capable of generating a modification of their environment, e.g., emit a sound through a speaker, grab with a robotic arm, or move a robot to a different position. Communicate actions are intended to send/receive data to/from one or more devices. Finally, Compute actions perform computations locally to the robot, e.g., averaging or shortest path.

To represent these activities, we need a compositional model to be able to compose and to abstract composition of actions to represent the collaborative goal of the system. An option can be a prescriptive modeling language such as BPMN and, in particular, the collaboration diagrams. Indeed, BPMN collaboration diagrams can be used to set and deploy distributed systems by applying model-driven engineering techniques as discussed in Chap. 8. BPMN collaborations represent the behavior of the single devices in the system and their interaction. Moreover, they are meant to be composed of different hierarchical levels that provide different views on the involved processes [13, 14]. Indeed, BPMN introduces process elements tailored to encapsulate fine-grained behavioral patterns, i.e., compositions of *task* elements into *subprocess* and *call activity* elements. Such elements are "*graphical object within a*

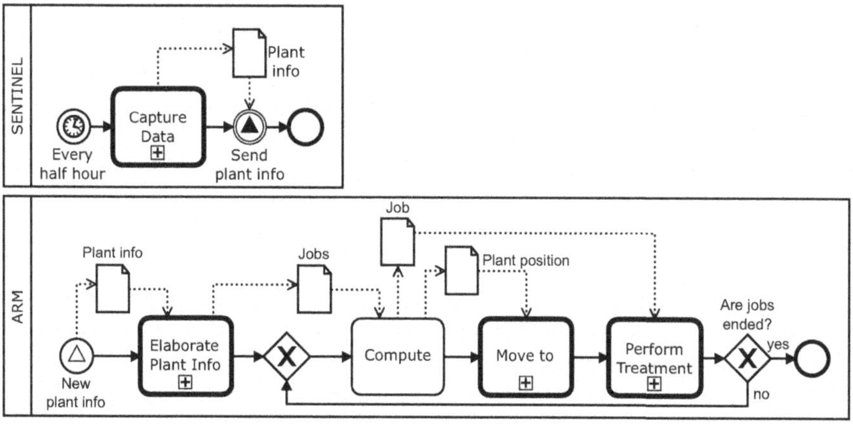

Fig. 9.3 Compositional model of the running example

Process, but it also can be opened up to show a lower-level Process" [33, Section 10.2.5, p. 173]. Therefore, BPMN collaboration supports hierarchical modeling, allowing to split part of the behavior from the goal of the main process. To sum up, concerning the structural dimension of the Bricks compositional model, a Brick is made of a task element that corresponds to a basic action performed by an IoT device. Multiple tasks can be then combined to form more complex activities and encapsulated in subprocesses or call activities.

Figure 9.3 depicts the BPMN collaboration that models the goal of the running example introduced in Sect. 9.2. The two pools contain both call activities (the squared elements with a thick border) that act as references to the actual behavior. Such mechanism guarantees either the composition of different computational units, i.e., Bricks, into more abstract representations, i.e., Bricks of Bricks.

For example, the call activity *Perform Treatment* performed by the robotic arm can be implemented in different ways depending on the environmental information gathered by the sentinel. Indeed, in case the sentinel spots the need of cutting leaves from the plant, the arm performs the behavior by the implementation of the *Perform Treatment* reported in Fig. 9.4a, while in case the sentinel spots the need of water, the arm performs the treatment depicted in Fig. 9.4b. Notably, the process of Fig. 9.4a contains other call activities that have multiple alternative implementation. For instance, the activity *Capture Image* can be done directly by the arm or delegated to the sentinel in case the arm's camera is not working.

9.4.2 *Causal Sense of Agency*

Agency is the capacity of an entity to learn and understand how its actions can affect the environment in which it lives [25]. One key aspect of Bricks is their capability

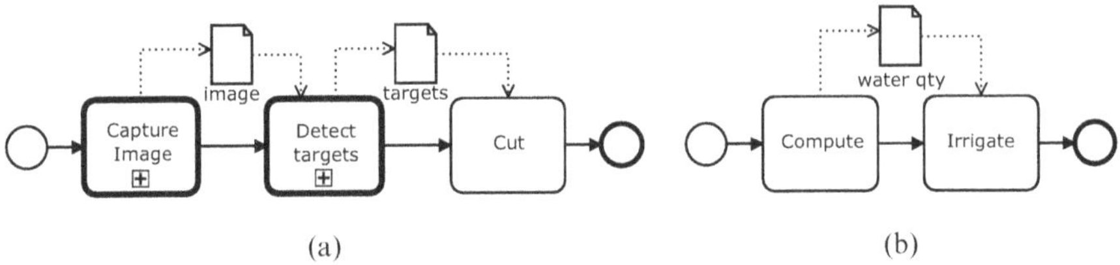

Fig. 9.4 Implementations of treatment call activity

to do so through environmental sensing and situation recognition, focusing on unpredictable changes. Through the perception of the Bricks' own abilities and how their enactment affects the runtime, an IoT system developed following the BRICKS principles should be automatically understood by the framework with no human intervention. Of course, this understanding will be different in different environments. Bricks understand their capabilities and role concerning their environment, namely, they build a model of the world and a model of themselves in the world.

Fully developed self-perception capabilities should extend to the recognition of how elementary Bricks' actions relate to the goal of the overall system; that is, let the system autonomously develop a sense of agency to identify automatically which dynamic reconfigurations can be enacted. A promising starting point for agency-enabled Bricks can be Bayesian networks.

9.4.3 Self-Integration

Self-integration is the capability of an autonomous system or process to integrate itself with other autonomous systems or processes. In BRICKS, methods and techniques will be investigated to compose Bricks into Bricks of Bricks, including a crucial evaluation step of the integrated capabilities to be executed into a sandboxed environment before injection into the runtime, preventing undesirable integrations to reach actual execution.

As part of the integration evaluation mechanism, a definition of how the functionality of Bricks is behaving (considering the context in which it is proposed) is necessary. Concretely, a realistic plan for such system to be developed is to require user guidance at first and then progressively isolate and reduce the need of such feedback up to the point of reaching full automation (possibly, depending on the safety-criticality of the context, limited to some conditions).

Testing integrations before performing them on the running system requires a sandbox. De facto, this instrument will realize a continuous integration process internal to the BRICKS platform. Since compositions need to be evaluated regardless of their source (human-made or automatic), the investigation about appropriate evaluation mechanisms can proceed in parallel with the work regarding self-integration. Besides evaluation, a paramount aspect of self-integration is the search for the most promising compositions: A "brute force" approach generating all possible composition will not be able to scale reasonably (as it grows combinatorially), so a search strategy is needed.

Once basic mechanisms of self-integration and evaluation are in place, the system should be able to support ways for a designer to express collaborative goals declaratively and having Bricks self-integrate to achieve them.

9.4.4 Overall Approach

To sum up, the overall approach starts with the modeling of the top-level BPMN collaboration, which defines the goal of the IoT system. Therefore, a step-by-step refinement of the devices' behavior is required to ensure to the devices alternatives for reaching the goal. The refinement terminates once the lower level of abstraction is reached and the different Bricks are present.

These models can be then enacted by the devices that should implement a sort of *augmented* process engines that on one hand orchestrate the execution of the activities and on the other hand, include mechanism of self-integration and self-perception.

More in detail, while the system is deployed and running, both developers and the system itself can propose new Bricks. While the manual development will be no different than traditional software development, entirely new components may be generated by the self-perception (agency) subsystem, based on the observations made on a sandbox. Once new Bricks are available, they can be combined into hierarchical structures both manually and automatically by the self-integration engine. Manual integrations are the result of a designer's decision and should carry an explanation for their purpose (typically, raise the abstraction level). Automatic self-integration proposals are the result of the evaluation of the self-integration engine and must be tested within a sandbox, typically implemented as a simulated environment. Once evaluated within the sandbox and verified as compatible with the overall goals, the most promising integrations can be proposed as replacements for the current ones.

9.4.4.1 Main Challenges

The runtime automated evolution via self-integration and agency brings new challenges that deserve specific attention:

- The *definition of the goals* must be expressed in such a way that it is possible to extract performance metrics both in the real system and in simulation.
- New integrations generated automatically should be *explainable*, and a promising technique could be the adoption of large language model.
- Combinations of Bricks must be *constrained* to elide configurations that are well-known to fail, and viable (and nonexclusive) options to achieve this objective

could be type systems, labeling systems, attachment of metadata, or rules defining nonsuitable options.
- The *proliferation of Bricks* from the self-agency system must be controlled, as it may lead to an explosion of possibilities that may be hard to manage.

9.5 Related Work

The results of BRICKS could significantly advance the state of the art. In particular, in learning a causal sense of agency, we exploit and extend recent results in these directions obtained for IoT systems [25]. The self-integration of Bricks could be dealt with by extending the aggregate computing platform [3]; the sandbox for evaluating Bricks compositions could exploit modern simulators such as Alchemist [35]. Finally, by adopting a process-driven view, we exploit a disciplined use of standard process notation [7, 12].

More in detail, our contribution is built upon the following three pillars.

Activity Recognition and Sense of Agency Situation recognition is an essential feature of IoT systems, enabling them to provide services tailored to emerging situations from the environment [40]. Typical examples are recognizing human activities to support assisted living in smart home environments [23] and identifying traffic jams to better regulate intersections [39]. Machine learning mechanisms can be leveraged to predict the likeliness of future situations and build services that exhibit anticipatory behavior [4]. However, these systems typically show a limited capacity to monitor the environment and the system response; BRICKS is instead designed with monitoring and explainability in mind. The need for systems to dynamically learn how to act in an environment (i.e., developing a sense of agency [25]) and to modify their behavior and structure in response to contingencies is attracting increasing attention in the areas of robotics [30] and reinforcement learning [24], calling for self-motivated approaches to learning [36] and causal reasoning [34]. BRICKS benefits from such recent findings, extending and adapting them to the specific characteristics of IoT systems.

Composable Approaches for Self-Adaptation The research for engineering methods and tools that enable the automatic integration of pieces of distributed computation to tackle a global goal is currently very active [1, 5, 21]. Several approaches in the literature take inspiration from natural processes, usually physics [28], (bio)chemistry [38], ethology [10], ecology [41], and biology [31]. Global goals are expressed implicitly in these systems, often emerging bottom-up from interactions among system components, and composition is horizontal. In BRICKS, global goals are first-class entities, and composition happens hierarchically, allowing (and promoting) top-down system design, composability at a larger scale, and evolution of the system behavior past its initial design. Another vital drive to the research in self-adaptation and integration comes from the multiagent systems, in which proactive autonomous components must often cooperate despite conflicting goals, and from collective adaptive systems, where the notion of a dynamic group is a first-class abstraction. Notable examples are Helena [20], DEECo [8], SCEL [32], Carma [26], and Aggregate Computing [2]; these approaches focus on clean, top-down engineering of composable distributed systems. Compared to them, in BRICKS, we introduce the ability to build behaviors through self-integration and test them within the infrastructure, creating a link between nature-inspired approaches and well-engineered distributed systems.

Augmented Process Management for IoT Bridging the gap between business processes and the IoT has been recognized as a rising need to support automation in cooperative scenarios [22]. The authors foresee their combination as a win–win situation [27]. Several attempts have been made at modeling IoT scenarios as processes [11], even in the presence of robotic devices [12, 15]. These formalisms model behaviors of IoT systems graphically, in the form of process-of-things and things-aware-processes [27], but struggle at providing timely reconfigurations under open and highly dynamic environments [6, 22]. Furthermore, we lack tools supporting adaptive IoT-aware systems [16, 19, 29, 37], where processes may reconfigure based on contingencies. In BRICKS, we aim to support the dynamic enactment of collaborative processes coordinating heterogeneous IoT systems exposed to unpredictable situations. The need for self-adaptation in the face of unforeseen contingencies in IoT-oriented middleware is a well-known issue in the literature [9, 19]. However, most of the existing proposals discover and instantiate variants of a system dynamically without acting on the overall business process [17]. The BRICKS approach is specifically meant to fill such a gap. We envision filling this gap by relying on the emerging concept of an augmented process management system that improves over the static and hard-coded classes of process-oriented systems by adapting to uncertainties.

9.6 Conclusion

In this chapter, we have presented the vision of BRICKS, a novel computational model for adaptable and open IoT systems to recognize contingencies and autonomously reconfigure their processes to achieve collaborative goals. To summarize, we can list the key innovations that BRICKS promises to bring about:

- The proposed BRICKS model goes beyond the state-of-the-art compositional models for distributed and collective systems, enabling contextual information to be built through an automated self-perception approach, thus introducing an entirely new way to deal with openness.
- BRICKS innovates by integrating ideas emerging sparsely in the literature into a coherent framework and pushing across the boundary by proposing an evolving and largely automated self-integration platform, thus fostering spontaneous adaptation to changes and contingencies of a dynamic and open environment.

- The definition of IoT systems in terms of augmented processes, along with the proposed support for self-perception and self-integration, will provide novel insights that we expect to be actionable to improve the current software engineering practices for the IoT. We also believe that the research in process-oriented engineering will benefit from the project's findings.

Currently, we are in the process of developing the presented ideas, in the hope of being able, with BRICKS, to actually deliver its identified potential.

References

1. 15th International Symposium on Software Engineering for Adaptive and Self-Managing Systems Symposium (2020)
2. Beal, J., Pianini, D., Viroli, M.: Aggregate programming for the Internet of Things. Computer **48**(9), 22–30 (2015). https://doi.org/10.1109/MC.2015.261
3. Beal, J., Pianini, D., Viroli, M.: Aggregate programming for the Internet of Things. Computer **48**(9), 22–30 (2015)
4. Becker, C., Julien, C., Lalanda, P., Zambonelli, F.: Pervasive computing middleware: current trends and emerging challenges. CCF Trans. Pervasive Comput. Interact. **1**(1), 10–23 (2019)
5. Bellman, K.L., Gruhl, C., Landauer, C., Tomforde, S.: Self-improving system integration - on a definition and characteristics of the challenge. In: IEEE 4th International Workshops on Foundations and Applications of Self* Systems, FAS*W@SASO/ICCAC 2019, Umea, Sweden, June 16–20, 2019, pp. 1–3. IEEE, Piscataway (2019)
6. Beverungen, D., Buijs, J.C., Becker, J., Di Ciccio, C., van der Aalst, W.M., Bartelheimer, C., vom Brocke, J., Comuzzi, M., Kraume, K., Leopold, H., et al.: Seven paradoxes of business process management in a hyper-connected world. Bus. Inform. Syst. Eng. **63**(2), 145–156 (2021)
7. Bourr, K., Corradini, F., Pettinari, S., Re, B., Rossi, L., Tiezzi, F.: Disciplined use of BPMN for mission modeling of multi-robot systems. In: Proceedings of the Forum at Practice of Enterprise Modeling. CEUR Workshop Proceedings, vol. 3045, pp. 1–10. CEUR-WS.org (2021)
8. Bures, T., Gerostathopoulos, I., Hnetynka, P., Keznikl, J., Kit, M., Plásil, F.: DEECO: an ensemble-based component system. In: Kruchten, P., Giannakopoulou, D., Tivoli, M. (eds.) CBSE'13, Proceedings of the 16th ACM SIGSOFT Symposium on Component Based Software Engineering, part of Comparch '13, Vancouver, BC, Canada, June 17–21, 2013, pp. 81–90. ACM, New York (2013)
9. Burzlaff, F., Bartelt, C.: A conceptual architecture for enabling future self-adaptive service systems. In: Bui, T. (ed.) 52nd Hawaii International Conference on System Sciences, HICSS 2019, Grand Wailea, Maui, Hawaii, USA, January 8–11, 2019, pp. 1–10. ScholarSpace (2019)
10. Charles, A., Menezes, R., Tolksdorf, R.: On the implementation of SwarmLinda. In: Yoo, S., Etzkorn, L.H. (eds.) Proceedings of the 42nd Annual Southeast Regional Conference, 2004, Huntsville, Alabama, USA, April 2–3, 2004, pp. 297–298. ACM, New York (2004)
11. Compagnucci, I., Corradini, F., Fornari, F., Polini, A., Re, B., Tiezzi, F.: A systematic literature review on IoT-aware business process modeling views, requirements and notations. Softw. Syst. Model. **22**(3), 969–1004 (2023)
12. Corradini, F., Pettinari, S., Re, B., Rossi, L., Tiezzi, F.: A BPMN-driven framework for multi-robot system development. Robot. Auton. Syst. **160**, 104322 (2023)
13. Corradini, F., Polini, A., Re, B., Rossi, L., Tiezzi, F.: Supporting multi-layer modeling in BPMN collaborations. In: Workshop on Enterprise and Organizational Modeling and Simulation. LNBIP, vol. 298, pp. 53–67. Springer, Berlin (2017)

14. Corradini, F., Polini, A., Re, B., Rossi, L., Tiezzi, F.: Consistent modelling of hierarchical BPMN collaborations. Bus. Process Manag. J. **28**(2), 442–460 (2022)
15. de Araújo Silva, E., Valentin, E., Carvalho, J.R.H., da Silva Barreto, R.: A survey of model driven engineering in robotics. J. Comput. Lang. **62**, 101021 (2021)
16. de la Croix, J.P., Lim, G.: Event-driven modeling and execution of robotic activities and contingencies in the Europa Lander mission concept using BPMN (2020)
17. Deng, S., Xiang, Z., Yin, J., Taheri, J., Zomaya, A.Y.: Composition-driven IoT service provisioning in distributed edges. IEEE Access **6**, 54258–54269 (2018)
18. Diaconescu, A., Frey, S., Müller-Schloer, C., Pitt, J., Tomforde, S.: Goal-oriented holonics for complex system (self-)integration: concepts and case studies. In: Cabri, G., Picard, G., Suri, N. (eds.) 10th IEEE International Conference on Self-Adaptive and Self-Organizing Systems, SASO 2016, Augsburg, Germany, September 12–16, 2016, pp. 100–109. IEEE Computer Society (2016). https://doi.org/10.1109/SASO.2016.16
19. Dumas, M., Fournier, F., Limonad, L., Marrella, A., Montali, M., Rehse, J., Accorsi, R., Calvanese, D., Giacomo, G.D., Fahland, D., Gal, A., Rosa, M.L., Völzer, H., Weber, I.: AI-augmented business process management systems: a research manifesto. ACM Trans. Manag. Inf. Syst. **14**(1), 11:1–11:19 (2023)
20. Hennicker, R., Klarl, A.: Foundations for ensemble modeling—the Helena approach—handling massively distributed systems with elaborate ensemble architectures. In: Iida, S., Meseguer, J., Ogata, K. (eds.) Specification, Algebra, and Software—Essays Dedicated to Kokichi Futatsugi. Lecture Notes in Computer Science, vol. 8373, pp. 359–381. Springer, Berlin (2014)
21. IEEE International Conference on Autonomic Computing and Self-Organizing Systems, ACSOS 2020, Washington, DC, USA, August 17–21, 2020. IEEE, Piscataway (2020)
22. Janiesch, C., Koschmider, A., Mecella, M., Weber, B., Burattin, A., Di Ciccio, C., Fortino, G., Gal, A., Kannengiesser, U., Leotta, F., et al.: The internet of things meets business process management: a manifesto. IEEE Syst. Man Cybern. Mag. **6**(4), 34–44 (2020)
23. Kim, E., Helal, S., Cook, D.J.: Human activity recognition and pattern discovery. IEEE Pervasive Comput. **9**(1), 48–53 (2010)
24. Leibo, J.Z., Hughes, E., Lanctot, M., Graepel, T.: Autocurricula and the emergence of innovation from social interaction: a manifesto for multi-agent intelligence research (2019). CoRR abs/1903.00742
25. Lippi, M., Mariani, S., Zambonelli, F.: Developing a "sense of agency" in IoT systems: preliminary experiments in a smart home scenario. In: 19th IEEE International Conference on Pervasive Computing and Communications Workshops and other Affiliated Events, PerCom Workshops 2021, Kassel, Germany, March 22–26, 2021, pp. 44–49. IEEE, Piscataway (2021). https://doi.org/10.1109/PERCOMWORKSHOPS51409.2021.9431003
26. Loreti, M., Hillston, J.: Modelling and analysis of collective adaptive systems with CARMA and its tools. In: Bernardo, M., Nicola, R.D., Hillston, J. (eds.) Formal Methods for the Quantitative Evaluation of Collective Adaptive Systems - 16th International School on Formal Methods for the Design of Computer, Communication, and Software Systems, SFM 2016, Bertinoro, Italy, June 20–24, 2016, Advanced Lectures. Lecture Notes in Computer Science, vol. 9700, pp. 83–119. Springer, Berlin (2016)
27. Maamar, Z., Kajan, E., Guidara, I., Moctar-M'Baba, L., Sellami, M.: Bridging the gap between business processes and IoT. In: Desai, B.C., Cho, W. (eds.) IDEAS 2020: 24th International Database Engineering & Applications Symposium, Seoul, Republic of Korea, August 12–14, 2020, pp. 2:1–2:10. ACM, New York (2020)
28. Mamei, M., Zambonelli, F.: Programming pervasive and mobile computing applications: The TOTA approach. ACM Trans. Softw. Eng. Methodol. **18**(4), 15:1–15:56 (2009)
29. Mass, J., Chang, C., Srirama, S.N.: Wiseware: a device-to-device-based business process management system for industrial internet of things. In: Liu, X., Qiu, T., Guo, B., Lu, K., Ning, Z., Dong, M., Li, Y. (eds.) 2016 IEEE International Conference on Internet of Things (iThings) and IEEE Green Computing and Communications (GreenCom) and IEEE Cyber, Physical and Social Computing (CPSCom) and IEEE Smart Data (SmartData), Chengdu, China, December 15–18, 2016, pp. 269–275. IEEE, Piscataway (2016)

30. Merrick, K.E.: Value systems for developmental cognitive robotics: a survey. Cogn. Syst. Res. **41**, 38–55 (2017)
31. Müller-Schloer, C.: Organic computing: on the feasibility of controlled emergence. In: Orailoglu, A., Chou, P.H., Eles, P., Jantsch, A. (eds.) Proceedings of the 2nd IEEE/ACM/IFIP International Conference on Hardware/Software Codesign and System Synthesis, CODES+ISSS 2004, Stockholm, Sweden, September 8–10, 2004, pp. 2–5. ACM, New York (2004)
32. Nicola, R.D., Loreti, M., Pugliese, R., Tiezzi, F.: A formal approach to autonomic systems programming: the SCEL language. ACM Trans. Auton. Adapt. Syst. **9**(2), 7:1–7:29 (2014)
33. OMG: Business Process Model and Notation (BPMN V 2.0) (2011)
34. Pearl, J.: The seven tools of causal inference, with reflections on machine learning. Commun. ACM **62**(3), 54–60 (2019)
35. Pianini, D., Montagna, S., Viroli, M.: Chemical-oriented simulation of computational systems with ALCHEMIST. J. Simul. **7**(3), 202–215 (2013)
36. Roohi, S., Takatalo, J., Guckelsberger, C., Hämäläinen, P.: Review of intrinsic motivation in simulation-based game testing. In: Mandryk, R.L., Hancock, M., Perry, M., Cox, A.L. (eds.) Proceedings of the 2018 CHI Conference on Human Factors in Computing Systems, CHI 2018, Montreal, QC, Canada, April 21–26, 2018, p. 347. ACM, New York (2018)
37. Teixeira, S., Agrizzi, B.A., Filho, J.G.P., Rossetto, S., de Lima Baldam, R.: Modeling and automatic code generation for wireless sensor network applications using model-driven or business process approaches: a systematic mapping study. J. Syst. Softw. **132**, 50–71 (2017)
38. Viroli, M., Zambonelli, F.: A biochemical approach to adaptive service ecosystems. Inf. Sci. **180**(10), 1876–1892 (2010)
39. Won, M.: Intelligent traffic monitoring systems for vehicle classification: a survey. IEEE Access **8**, 73340–73358 (2020)
40. Ye, J., Dobson, S., McKeever, S.: Situation identification techniques in pervasive computing: a review. Pervasive Mob. Comput. **8**(1), 36–66 (2012)
41. Zambonelli, F., Omicini, A., Anzengruber, B., Castelli, G., Angelis, F.L.D., Serugendo, G.D.M., Dobson, S.A., Fernandez-Marquez, J.L., Ferscha, A., Mamei, M., Mariani, S., Molesini, A., Montagna, S., Nieminen, J., Pianini, D., Risoldi, M., Rosi, A., Stevenson, G., Viroli, M., Ye, J.: Developing pervasive multi-agent systems with nature-inspired coordination. Pervasive Mob. Comput. **17**, 236–252 (2015)

Chapter 10
Learning Opportunities in Collective Adaptive Systems

Gianluca Aguzzi, Roberto Casadei, Stefano Mariani, Mirko Viroli, and Franco Zambonelli

10.1 Introduction

A Fluidware system is a specific example of Collective Adaptive Systems (CAS). Adapting the definition from [8], CAS are distributed systems featuring multiple computational entities (e.g., agents) that do not have global knowledge about the system state. To achieve collective goals, these entities must interact with each other to either avoid interference or help each other, as they need information beyond their local observation capabilities.

Within a Fluidware system, there exist three key constituents: the *funnel process* abstraction, the Fluidware middleware, and a computing infrastructure (e.g., a cloud–edge system). At the core of this framework resides the notion of the funnel process—a versatile abstraction crucial for crafting collaborative applications. This funnel process operates as both a procedural entity and an information flow consumer, abstracting the complexities associated with individual devices. Its capabilities encompass capturing, merging, splitting, and replicating these information flows, thereby facilitating seamless manipulations across spatial and temporal dimensions throughout the entire infrastructure. Concurrently, the middleware works as the carrier of the complete collective application logic. Specifically, it undertakes the orchestration of funnel placements within the edge–cloud continuum, scheduling the computations at opportune nodes, and the coordination of data movement among diverse funnels.

G. Aguzzi (✉) · R. Casadei · M. Viroli
Università degli Studi di Bologna, Bologna, Cesena, Italy
e-mail: gianluca.aguzzi@unibo.it; roby.casadei@unibo.it; mirko.viroli@unibo.it

S. Mariani · F. Zambonelli
Università di Modena e Reggio Emilia, Reggio Emilia, Italy
e-mail: stefano.mariani@unimore.it; franco.zambonelli@unimore.it

F. Zambonelli et al. (eds.), *Fluidware*, Internet of Things,
https://doi.org/10.1007/978-3-031-62146-8_10

Adaptation within the Fluidware system manifests at two pivotal levels: at the application level of funnel processes and at the middleware level. At the application level, adaptation pertains to the dynamic adjustment of individual funnel processes to align with changing environmental conditions. For instance, consider a scenario where a Fluidware system is deployed in a smart building. As occupancy patterns shift throughout the day, the funnel processes responsible for collecting and processing sensor data from various rooms adapt their parameters to optimize energy consumption, maintaining a comfortable environment while minimizing wastage. On the other hand, adaptation at the middleware level involves the reconfiguration of the Fluidware middleware to accommodate shifts in system requirements or resource availability. Take the case of a Fluidware system used for traffic management in a smart city context. During periods of heavy traffic congestion, the middleware might dynamically allocate funnel processes to elaborate and analyze incoming data streams from traffic sensors in real time. As traffic conditions normalize, the middleware could scale down these resources to avoid unnecessary overhead. However, achieving *effective* adaptation in such systems is complex due to the highly dynamic and unpredictable environment in which each agent operates. Centralized or hierarchical adaptation is often not feasible, as it depends on unrealistic assumptions such as the ability of a single controller to obtain a consistent and complete system-wide view and to process this information, reason, and distribute decisions in time. Instead, recent research has demonstrated that *decentralized self-adaptation* or *self-organization* can overcome issues associated to centralized adaptation. This paradigm shift increases resilience, improves scalability, and eliminates single points of failure.

To address these challenges, agents could be enhanced with *learning* capabilities, allowing them to instantiate and refine learning models using the knowledge acquired from observing their environment and peers. These learning models provide the means to improve quality attributes such as performance and cost for an individual agent or the entire collective. However, learning in CAS is challenging since there are several levels that need to be considered, starting from how to *deploy* a concrete collective application, how the middleware should execute it *efficiently*, there including communication and coordination overhead, and how the application can be engineered. Furthermore, CAS are obviously engineered by software designers and developers; hence learning could also be used to assist them at design time.

Accordingly, in this chapter, we first briefly overview the state of the art in CAS learning (Sect. 10.2), then we describe a few novel approaches developed within the Fluidware project to push forward its vision (Sects. 10.3, 10.4, 10.5, and 10.6), and finally we discuss further opportunities to plug learning into the Fluidware vision (Sect. 10.7).

10.2 State of the Art in Machine Learning for Collective Adaptive Systems

In recent years, there has been significant development in the field of machine learning techniques, in both supervised (e.g., AlexNet and similar) and unsupervised (LLM) modes, as well as reinforcement learning (RL) and their combinations (e.g., ChatGPT). Among the many techniques, multiagent reinforcement learning (MARL) is one of the most promising: It is a subfield of RL that focuses on studying the behavior of multiple learning agents that coexist in a shared environment. Each agent is motivated by its own rewards and does actions to advance its own interests; in some environments, these interests are aligned with the interests of other agents, resulting in *cooperative* or *collective* behavior. MARL is a crucial research area that aims to develop robust, scalable, and resilient multiagent systems. Numerous approaches, algorithms, and solutions have been proposed in the literature over the years to address various challenges and opportunities in MARL. However, MARL has a wide range of applications in different domains, which motivates different perspectives and classifications of MARL algorithms. Therefore, several surveys have been conducted to provide comprehensive overviews of specific aspects or categories of MARL algorithms. In the following, we present the relevant surveys in the context of Fluidware applications:

1. *On Learning in Collective Self-Adaptive Systems: State of Practice and a 3D Framework [8]*—In this survey, the authors focus on learning in the context of collective adaptive systems. They identify the main algorithms used (reinforcement learning, supervised learning, unsupervised learning) and create a framework to classify different applications with respect to autonomy (full vs. restricted), behavior (selfish vs. altruistic), and knowledge access (maximal vs. minimal).
2. *A Survey of Self-Organization Mechanisms in Multiagent Systems [48]*—This survey explores how self-organization is accomplished and used in CAS. The authors highlight how self-organization can assist reinforcement learning to converge to good behaviors, reducing the problem of concurrent learning in MAS/CAS.
3. *Multiagent Reinforcement Learning: A Selective Overview of Theories and Algorithms [51]*—This recent survey delves into MARL settings for both competitive and cooperative tasks, focusing on the theoretical background. It also discusses network and many-agent settings, with an emphasis on mean-field solutions and their convergence guarantees.
4. *Decentralized Multiagent Reinforcement Learning with Networked Agents—Recent Advances [52]*—This survey primarily focuses on network MARL studies in both cooperative and competitive settings. It does not cover CTDE (centralized training and decentralized execution) as learning in such scenarios should be performed online without a central authority to aggregate the system view and improve shared policies. The scale is not the primary concern here,

but rather the decentralization. In Fluidware, scale becomes important as we generally cannot know the whole population of the system.

5. *A Survey and Critique of Multiagent Deep Reinforcement Learning [18]*—In this chapter, the authors concentrate on deep learning aspects, highlighting the advantages of MARL in various settings, including cooperative, competitive, and mixed scenarios. They categorize the main classes as Analysis of Emergent Behavior, Learn to Communicate, and Agents Modeling Agents. While they do not primarily focus on many-agent aspects, they do discuss them in the context of few agents.
6. *A Review of Cooperative Multiagent Deep Reinforcement Learning [30]*—This chapter solely focuses on the cooperative aspect of deep MARL. The algorithms are classified based on how agents learn, leading to independent learners, full observable critics, value decomposition functions, and consensus methods. While the main goal is not large-scale scenarios, they also discuss some work related to many-agent approaches like mean field.
7. *A Survey of Multiagent Reinforcement Learning with Communication [53]*—This survey mainly centers on communication in the context of MARL. It discusses the design perspective, including the learning scheme, agent communication, channel construction between agents, and policy development, among other approaches discussed in the literature.
8. *Multiagent Deep Reinforcement Learning: A Survey [15]*—Another survey focusing on the MARL perspective but with a different classification. It first discusses the learning schemes (Centralized, Decentralized, Centralized Training, and Centralized Execution) and addresses the challenges of the current landscape as of 2019.
9. *Cooperative Multiagent Learning: The State of the Art [31]*—An older reference of MARL in the context of cooperative tasks. The survey explores different types of policies (homogeneous vs. heterogeneous) and various ways of learning (team-based vs. concurrent) in cooperative multiagent settings.
10. *Multiagent systems: A survey from a machine learning perspective [43]*—Another older but seminal reference, actually one of the first to consider learning in MAS. It categorizes how learning affects MAS under different criteria (homogeneous vs. heterogeneous agents, communication vs. no communication).

Focusing solely on CAS, learning can be applied at different major levels: (i) at the *application* level (i.e., inside the funnel processes), (ii) at the *middleware* level, and (iii) at the *deployment* level [2]. Also, it can be used at design or run time, for either supporting CAS developers and administrators or directly injecting adaptiveness in the distributed components. At the application level, learning is used directly to solve collective tasks. For example, in the context of swarm robotics, it involves the ability of a group of agents to form a flock [42]. At the middleware level, the focus is on solving nonfunctional aspects. This includes reducing distributed computing consumption or minimizing the number of exchanged messages. At the deployment level, research focuses on how to opportunistically move computation in complex network infrastructures, such as cloud–edge–fog environments.

Reinforcement learning is extensively utilized at all levels in these systems due to its flexibility, online adaptability, and scalability. The focus of this chapter is mainly on reinforcement learning, and in particular multiagent reinforcement learning, as it naturally embraces a *distributed collective* of learning agents, while considering their strict relationship with the environment—as collective adaptive systems are. Closely related to reinforcement learning is another form of learning that recently got increasing attention from the RL and MARL communities: *causal learning* [17]. Causal learning is a discipline that studies how to learn causal models of an environment or phenomenon of interest through *interventions* [38], that is, through interaction with such an environment (or, phenomenon)—as (MA)RL. The main difference with respect to RL—as well as other forms of statistical machine learning, such as supervised and unsupervised learning—is that causal learning is able to tell apart cause–effect relationships between variables from mere associations: Traditional ML cannot give this guarantee by construction (see work from Judea Pearl and others, such as [40]). Section 10.5 gives an example.

10.3 Learning Along Funnel Processes

In Fluidware systems, the application logic is defined as a collective program, which is then executed by the middleware. The way in which this program is defined is crucial for the overall performance of the system. Among the many approaches, macroprogramming is one of the most promising: The focus is switched from the individual agent to the collective, having a top-down global-to-local approach and having the ability to combine simple behaviors to create complex ones—as exemplified by the aggregate computing paradigm [45]. However, in practice, constructing these aggregate computing applications is anything but straightforward. This is because, even if the paradigm supports adaptive computation, the programmers must have a deep understanding of the environment in which the system will be deployed. Indeed, certain basic components/behaviors work well in specific environmental dynamics but not in others, thus leaving the burden on the programmer to choose the most suitable combinations of blocks for a given environment. To address this issue, modern approaches focus on combining parts of macroprogramming paradigm with machine learning techniques. The idea is to preserve the characteristics of macroprogramming, such as declarativeness and composability, while incorporating those of machine learning, including adaptability and the ability to solve complex programs through "by-doing" or "by-example" methods.

There are various ways in which this combination can be achieved. For instance, in [3], "pieces" of a partially specified program can be filled in, resembling the field of program synthesis. On the other hand, in [4], the result of the macro program is used to guide the learning process, enabling more agile synthesis of the collective controller (Fig. 10.1 gives an overview of both ideas).

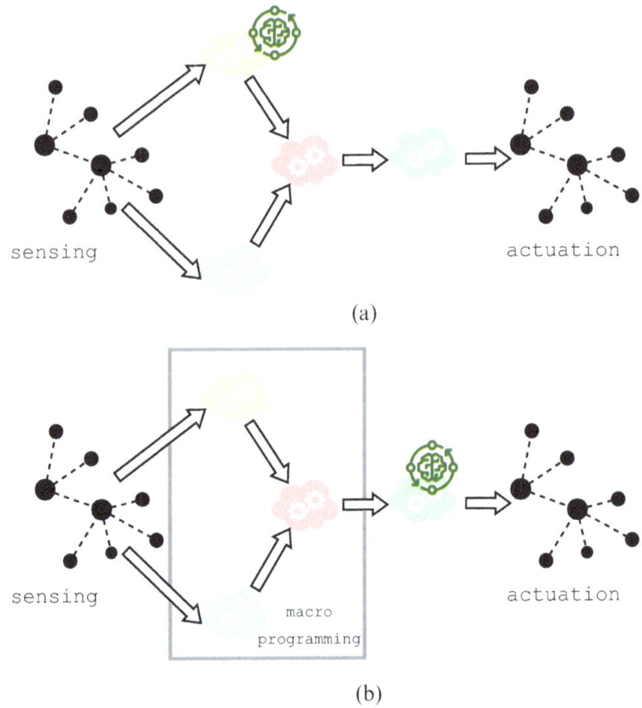

Fig. 10.1 High-level overview of the two approaches. (**a**) Funnel synthesis. (**b**) Program learning

Focusing on the foster approach, giving a complex network of funnels devising with macroprogramming, the idea is to learn only the parts of the whole network that are: i) hard to define, since they are inherently complex, or ii) strictly dependent on the environment in which the system is deployed. Therefore, the programmer will leave some funnel processes undefined, and the learning algorithm will fill in the gaps, in a process that is similar to program *synthesis/sketching*. In [3], the authors propose a variant of Q-learning that can be used to synthesize collective distributed programs basing part of the program on aggregate computing paradigm. Particularly, the learning process is performed at simulation time, where the agent collects *local* experiences in a *global* reply buffer. During the learning process, the agent executes the program as usual, but when it reaches a "sketching" point, it will execute a random action with a certain probability; otherwise it will execute the action suggested by the Q-table. The Q-function is shared and updated globally, but agents can only observe their local state and the messages received from their neighbors. That means that once the Q-function is learned, each agent can execute the program locally, without the need of a central authority. The buffer is then used to refine a Q-table with the usual Q-learning algorithm. This training procedure is known as centralized training and decentralized execution (CTDE) [13] (Fig. 10.2a).

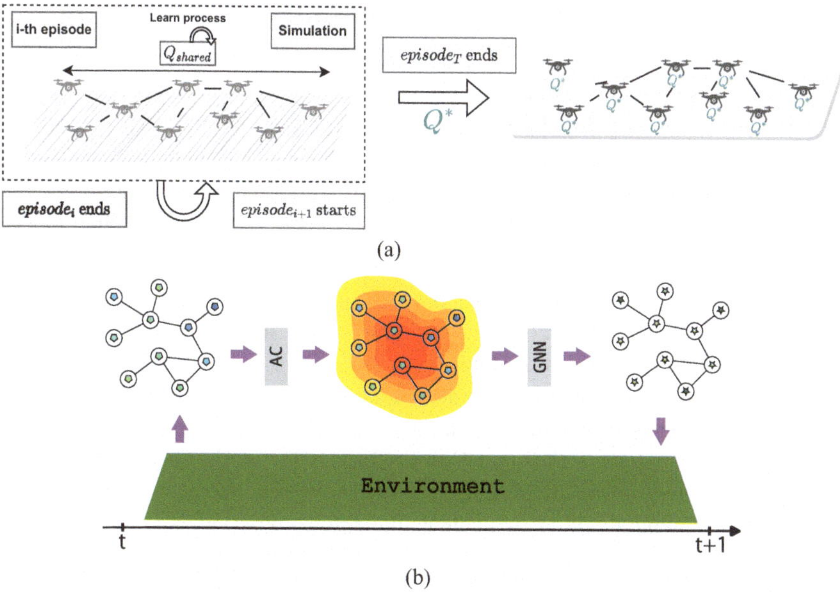

Fig. 10.2 Overview of the two approaches. (**a**) Overview of the training procedure. (**b**) Overview of field-informed reinforcement learning

The approach is then used to successfully learn a correction policy in order to speed up the convergence of a self-stabilizing algorithm.

In [4], the authors propose a different approach, called *Field-informed Reinforcement Learning* (FiRL), where the program computed by the macroprogramming paradigm is used to guide the learning process. Particularly, the program is used to enrich the state space of the learning algorithm with collective feature, ease up the learning process, and improve the quality of the learned policy (Fig. 10.2b). Indeed, in MARL algorithm the communication is often implicit, and the agents must learn how to communicate with each other to achieve the desired goal. Adding that information to the state space, the agents do not need to learn how to communicate, but only to leverage the information provided by the program. FiRL is based on Deep Q-learning, where the Q-function is approximated by a *graph neural network* (GNN). The choice of the GNN was motivated by the fact that it generalizes well to unseen topologies and can be used locally, even if it can be trained from the whole graph during the training phase. This combination of macroprogramming and learning is used to learn a collective controller for a swarm of robots, where the goal is to cover a moving phenomenon (e.g., a wildfire) that is not known a priori and spans a large area, meaning that no single agent can cover it alone and can detect the whole phenomenon.

10.4 Learning Scheduling Policies to Optimize Middleware

A crucial aspect for the Fluidware middleware is the scheduling of the funnel processes, and this is particularly import in macroprogramming, as it determines how the collective program is executed. Some of these macro approaches, such as Buzz [34] and aggregate computing [45], employ a *proactive* execution model that breaks down distributed execution into rounds typically composed of three macro phases:

1. *Creation of an evaluation context*: In this phase, the necessary information is gathered to evaluate the collective program (e.g., messages from neighbors, sensory information, previous state).
2. *Program evaluation*: Given the context, each node evaluates the defined collective program, creating the desired output.
3. *Actuation and message sharing*: With the produced output, nodes can perform actions that modify the environment. Furthermore, the necessary messages are exchanged among neighbors to evaluate the defined program.

The various models do not specify "when" these phases should occur, thereby leaving some decision-making margin to the middleware. In fact, tuning the evaluation frequency can impact various aspects, including energy consumption and convergence times to the defined collective structure.

One of the earliest works focused on programming distributed schedulers [5], where rules expressed with aggregate programming allowed the description of schedulers that react to environmental/temporal events. However, despite the interest in this approach, there was the problem of having to program these behaviors in advance, with the disadvantage of potentially overlooking unrecognized environmental dynamics and making the controller definition hard.

Therefore, in [1], an attempt was made to create these distributed schedulers through reinforcement learning—particularly a variant of Q-learning—with the goal of optimizing a certain family of distributed computations, namely *self-stabilizing* computations—that are programs that have the characteristic of achieving a stable output given a stable context. In the proposed approach, agents are equipped with a Q-table that is used to decide when to evaluate the collective program, i.e., the action space consists of the next wake-up time. As in the previous work, the agents collect local experiences in a global reply buffer, and the Q-table is updated using the CTDE paradigm. The simulations show that the learning algorithm is able to reduce the energy consumption of the system, while maintaining the same convergence time of the self-stabilizing algorithm.

10.5 Learning the Environment Dynamics

In scenarios where multiple sensor and actuator devices have many interdependencies, it is impractical (if not impossible) for the system designer to foresee (i) all the possible relations between the environmental variables and (ii) all the possible effects of acting on them. Hence, in a Fluidware perspective, it would be very difficult to design funnel processes with guarantees to produce desired results (e.g., in terms of affecting the environment in a desired way). Yet, for funnel processes and software agents alike, to have explicit knowledge of the network of cause–effect relations (aka the causal model of the agents–environment system [32]) is of fundamental importance to make sense of their operational environment, autonomously decide plans of actions, and being able to explain such decisions [6, 25]. Thus, having software agents (such as funnel processes) autonomously *learn* such causal model once deployed in the target (simulated) environment, with minimal a priori knowledge, may be extremely valuable.

Causal learning techniques [16] enable distinguishing causal relations from mere associations. For instance, recognizing that the states of the air conditioning system and of the temperature in a room are not simply correlated, but that the first causes the second to change. The result of such causal learning process is thus a *causal network* relating causes and effects with directed edges, a task that has been shown to be out of reach for statistical machine learning models [40], which rely on purely *observational* data to learn *associations*. It is worth noting here that also funnel process defines a network of processing steps, in a sense, hence can be expressed in terms of causal graphs: Some nodes will be variables, and some will be operators manipulating such variables. Learning a causal model would then correspond to automatically assembling funnel processes.

In most existing proposals, though, causal learning assumes a *centralized* setting, with a single learner/agent having full observability and control over the environment variables, or with a central site aggregating distributed observations or merging local models [16, 17, 46]. However, the centralized assumption can rarely hold for many real-world deployments, and especially for CAS, where multiple agents are typically deployed in different regions of an environment. There, each agent only has partial observability of the environmental variables and the capability of affecting only a limited portion of the overall environmental variables through its own actuators (Fig. 10.3). Hence, authors of [28] propose a *multiagent protocol for distributed learning of causal networks*, aimed at letting agents in an MAS (but could be CAS, or a Fluidware system with funnels instead of agents) cooperate in unveiling causal relationships that individuals could not reveal by themselves, due to partial observability.

They consider N agents $\mathcal{A} = \{\mathcal{A}_i\},\ i = 1, \ldots, N$ willing to learn a causal network M_i of the relationships between V environment variables. The set V is partitioned into two (possibly, empty) sets, one of *controlled* variables C (i.e., corresponding to actuators) and one of *uncontrolled* variables U (i.e., sensors), such that $V = C \cup U$. It is assumed that each agent knows an algorithm for *independently* learning its own local causal network $\mathcal{L}_i$, that is, the causal network learned by relying solely on its own known variables P_i (i.e., without running our multiagent protocol). It does not really matter which one, but they choose one already adopted in an agent-oriented scenario [12]. Each agent has *partial observability of the environment*, that is, is aware of only P_i out of the V variables. Formally: $P_i = C_i \cup U_i,\ P_i \subset V \Rightarrow C_i \subset C, U_i \subset U$. However, no variable in V

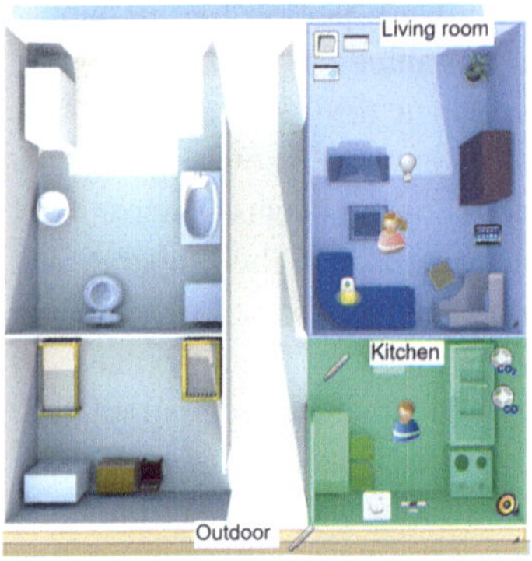

(a)

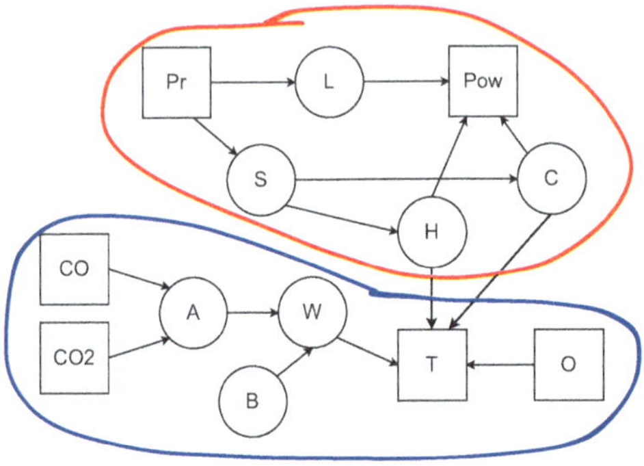

(b)

Fig. 10.3 Partial observability in CAS. (**a**) Two agents with limited observability. (**b**) The ground truth causal network

can be unknown to every agent, that is, $V = \bigcup_{i=1}^{N} P_i$—note that an overlap between the different P_i is admitted. M_i is not necessarily the *global* causal network G, that is, the one including every variable in V and every link among them. Rather, it is the "minimal" causal network that each agent "needs" to correctly understand situations and plan actions. That is, M_i contains all the variables in P_i and their links, plus the subset of variables in $V \setminus P_i$ that have links with variables in P_i.

The proposed protocol revolves around agents exchanging observational data and carrying out *interventions* whenever possible, that is, changing the values of their controlled variables through the simulation software that provides access to virtual devices (iCasa is used [22]). For instance, an agent may repeatedly (at a given time interval, during a given time period) activate and deactivate the A/C to check (or let other agents check) whether the temperature changes as a consequence. Interventions are the fundamental mechanisms to check whether two variables are linked in a cause–effect relationship and not by a mere correlation [33]. Intuitively, an intervention deliberately changes the value of a variable, *all others being untouched*, to see whether it affects others. As such, interventions require the ability to *control* variables, as is the case of the variables expressing the status of actuators in agent systems.

The authors show that multiagent causal learning has superior accuracy with respect to the single-agent case, as it can learn more causal relationships, while doing less mistakes (causal links learned but not present in the ground truth)—Fig. 10.4.

In Fluidware, causal graphs such as the one shown in Fig. 10.4 may guide funnel processes' own computations. On the one hand, they can use the graph to decide where to pick data from to make a particular decision (e.g., check CO levels to decide what to do with the window W). On the other hand, the same graph can also be used to predict outcomes of actions ("what if we close the window?"). In

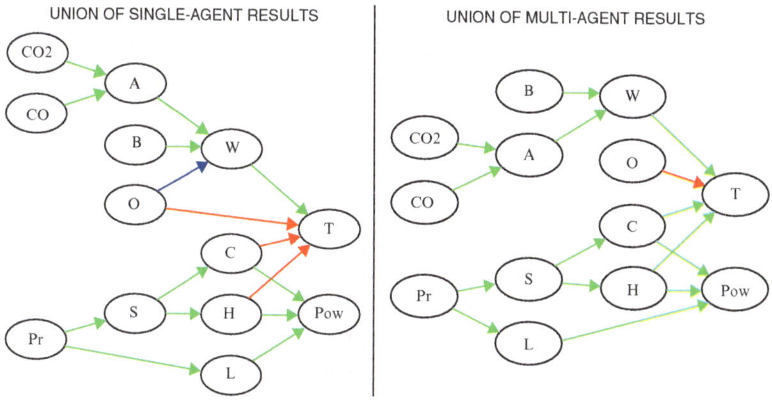

Fig. 10.4 Multiagent learning outperforms single-agent causal discovery: It makes less mistakes (red arrows) and finds less false causal relations (blue arrows)—green arrows are correct

summary, causal graphs are a flexible and powerful tool to make decisions, and learning them is a great enabler of adaptation (if the environment changes, a new graph can be learned to cope with it).

10.6 Learning Local Communication for Collective Goal Achievement

In CAS where agents' observability of each other actions and their effects has no limits, coordination may be implicit [23]: Each agent reacts to what others are doing as good as it can. In more challenging scenarios, however, *explicit* communicative actions are needed: Agents need to tell each other what to do. Such communication actions are often assumed to be among the built-in capabilities of agents, and the communication policies resulting from their composition are defined according to the specific goal to achieve. However, *learning* such policies would enable agents to adapt to new deployments, run-time changes, and/or changing goals with minimal designers' intervention [24, 35]. The same applies to Fluidware, at multiple levels: At the application level, different funnel processes with different functions can be composed (akin to service composition) to achieve the system goals, and such a composition may require communication between them, that could be learned; at the middleware level, handling distributed event flows to feed funnel processes as well as their composition requires processing nodes to communicate, and again the employed protocol could be learned.

Multiagent Reinforcement Learning (MARL) provides methods for learning to communicate, either as a form of language (ungrounded communication) or with practical environmental actions (grounded communications). Stigmergic communication via *pheromone* is a kind of grounded communication: A biological marker is deposited to persist for some time and diffuse within some range (thus can be sensed by others locally in time and space). Such a mechanism is pervasively used in natural systems [19], in MAS, and swarm robotics as well [20, 27, 50] for successful self-organized behaviors to emerge. Ant foraging is an example, with the Ant Colony Optimization algorithm for distributed optimization [11], while termites' nest building [10] and slime mold aggregation [39], which we use as our scenario below, are others. Slime mold cells [37] (the agents) have innate policies ("rules") to deposit pheromone (i.e., *when*, in what circumstances) and to follow it (e.g., instead of wandering randomly) with the individual goal to stay close to as many other slime mold cells as possible, contributing to the global one of assembling clusters—through self-organization.

The literature reports an attempt at learning such policies from scratch, that is, *when* to deposit and follow pheromone, whereas most of the related literature in related fields, e.g., evolutionary computation, deal instead with *how much* pheromone to deposit. The authors of [29], in fact, adopt independent Q-learning in a slime mold cells MARL environment where agents have partial observability (e.g., they perceive communications within a very limited range) and achieve a number of results (see Fig. 10.5 as reference). First, pheromone-based communication is learned consistently despite nonstationarity; second, the learned policies are quicker to converge to clusters than the built-in rules usually adopted in artificial swarms; third, agents do learn to self-organize to best achieve the ascribed systemic goal, by coordinating action selection so that pheromone deposit happens in a few places (to have a few bigger clusters) and persists in time (to not disperse them); finally, learning to communicate is also able to adapt to different collective goals, as agents do adapt by learning new policies with minimal external intervention (e.g., by solely changing the reward function).

Figure 10.5 shows an example run: In later stages of training, agents (red triangles) have learned how to exploit pheromone-based communication to achieve the clustering goal. In fact, agents have learned how to collectively perform the correct actions by distributing with the right balance the actions to make, in three fundamental ways: (1) Few deposit pheromones steadily while not moving at all, to communicate to others their presence. (2) Most of the agents look around searching for pheromone, that is, are eager to listen to "messages" (pheromone deposit) coming from others, so as to promptly move in their direction. (3) Another few still roam randomly, ignoring clustering messages (pheromone trails) to check whether better clustering opportunities exist. Without (1), for instance, if too many agents spam pheromone, it may happen that agents try to cluster in too many locations. Without (2), that is, if too few agents react to pheromone trails, clusters will not reinforce enough and may never form. Without (3), it would be impossible for behaviors like the one depicted in Fig. 10.5 to happen: Merging clusters to form a bigger one would be very difficult as agents would always prefer gravitating around existing ones.

In the context of Fluidware, these results demonstrate that it is possible and effective to learn communication policies (or, composition alike) even in CAS with many agents, hence despite nonstationarity. By learning communication policies at

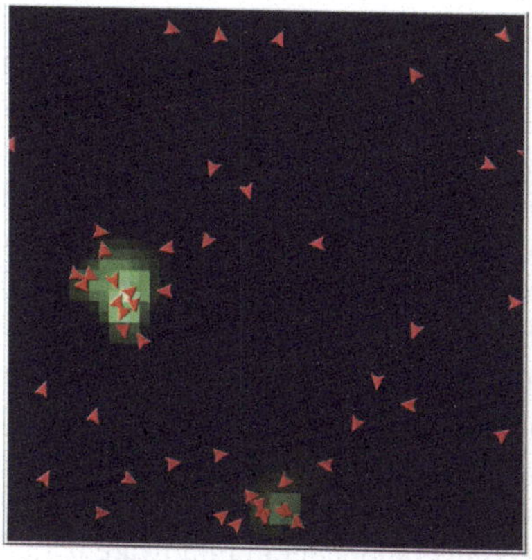

(a) Early on in the episode.

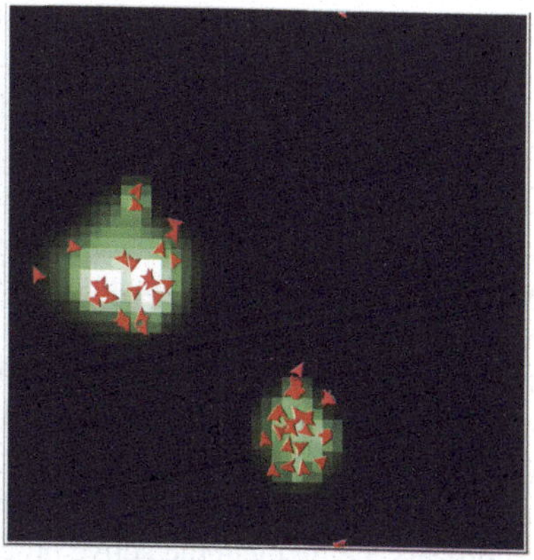

(b) Later, same episode.

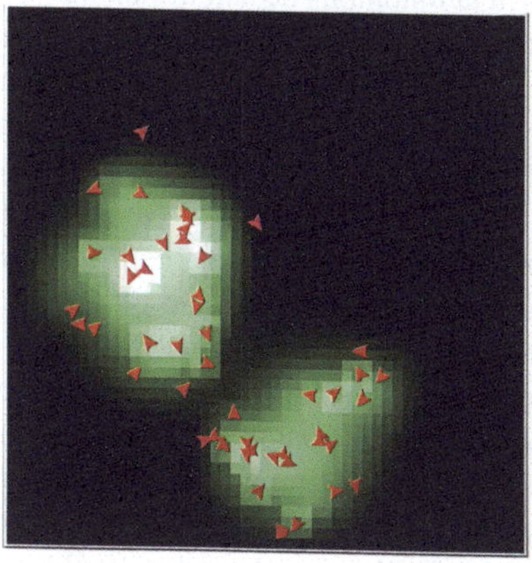

(c) Still same episode, later on.

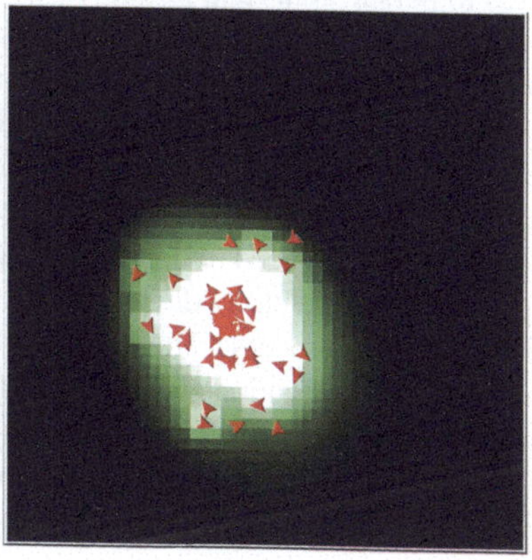

(d) End of episode.

Fig. 10.5 Agents learn a collectively effective distribution of communicative actions: (**a**) few spam pheromone (to form a few clusters), (**b**, **d**) most follow pheromone trails (to aggregate), (**c**) few persist to roam randomly (to take the chance of merging clusters)

run time, with MARL, it would be possible to adapt to contingencies both at the middleware level (e.g., broken communication links or delayed nodes) and at the application level (e.g., funnel processes reconfiguring their composition for a new goal to pursue).

10.7 Collective Learning and Fluidware

The key innovative idea of Fluidware is to abstract collectives of devices of the IoT fabric such as *sources, digesters, and targets* of *distributed "flows" of contextualized events*, carrying information about data produced and actuating commands. Accordingly, programming services and applications implies *declaratively specifying "funnel processes" to channel, elaborate, and redirect such flows* in a fully distributed way, as a means to coordinate the activities of devices and realize services and applications.

This overview already gives clues about where learning could be applied within Fluidware based on its cornerstone abstractions and mechanisms. First, Fluidware decouples devices from event/data streams; hence learning what device should be feeding or consuming which stream is an opportunity to consider. Second, such streams must be connected to devices and processing nodes acting as prosumers of events: How to connect them (in what topology, with what communication protocol, etc.) is another potential target for learning. Finally, and perhaps most naturally, learning can be utilized to synthesize, refine, or adapt the "behaviour" of funnel process, being them the active processing component in Fluidware paradigm devoted to pursue goals and possibly perform autonomous decision-making.

Besides these, other opportunities for learning may be devised out by considering the application scenarios that Fluidware aims to tackle, and their typical challenges and requirements. The potential high number and density of *deployed devices* raises issues about communication topology and fault tolerance: What if devices move? What if network links break? Finding the best way to communicate may be a task apt for learning. The need to operate on top of a dynamically *evolving infrastructure* comprising IoT/edge/cloud devices and resources raises deployment issues: Where to place running process? How to best relocate them at run time to cope with processing load fluctuations or hardware failures? Such a deployment policies may be another likely target for learning. Finally, the need for goal-oriented *orchestration* of distributed activities across devices of heterogeneous computational power (from tiny devices to cloud servers) raises issues about coordination and cooperation of distributed autonomous components for which MARL is particularly suited for.

The following sections elaborate on these opportunities.

10.7.1 Funnel Processes That Learn

Funnel processes are, essentially, the "algorithms" that the application designer defines and the Fluidware platform runs to achieve the application goals. Then, a first learning opportunity comes from *program synthesis* with the *sketching* technique [41] also suggested in [2] and experimented in [1]: Ask the developer to only specify a program "blueprint," e.g., a loose specification of the "high-level" expected behavior with missing lower-level details, that will be completed through learning—e.g., action policies learned by reinforcement learning, or data acquisition operations learned via prediction models.

Another learning opportunity comes when considering that funnel processes may be *composed* to modularly build more complex functionalities on top of simpler ones. Such compositions may explicitly design or learn [36] an increasingly adopted approach to deal with the complexity and uncertainty on dynamic software environments. In [26], for instance, reinforcement learning is used in a multiagent setting to learn the most appropriate matching between requests for services and services providers, also considering how to compose existing services in novel ways when a readily available service cannot meet the service request's requirements.

However, service composition does not account for any arbitrary *coordination* of funnel processes toward achievement of complex system goals. For instance, it may be the case that access to a given events flow must be given to a single funnel process at a time; hence a distributed mutual exclusion protocol must be enacted to coordinate competing funnels. Or, funnel processes may need to learn how to coordinate on the fly to perform adaptation actions in face of unexpected system disruption. Learning to coordinate, either implicitly or via explicit communication, is an open challenge that has seen increasing attention in recent years, especially in the multiagent reinforcement learning community [7, 44, 49].

10.7.2 Learning to Wire Event Flows to Funnels

Funnel processes work off of event flows, that is, by consuming and producing unbounded streams of events stemming from the data emitted by deployed devices. Such event flows are decoupled from the devices that produce them, from the standpoint of funnel processes (and their programmers). That is, they do not care which device contributed to which event flow, and how they only care about having access to the right event at the right time. This means that at different points in time the same funnel process may actually be connected (behind the scenes) to different devices, since different devices may be feeding the same "logical" event flow—depending on run-time contingencies. Learning how to wire this connection dynamically is an interesting opportunity.

For instance, devices may produce measurements with different accuracies and confidence in their values over time, due to battery issues, signal degradation, or any other reasons. A funnel process in charge of a critical task may require events always with maximum accuracy and confidence; hence predicting a loss of accuracy may be crucial to dynamically rewire devices to event flows.

Also, funnel processes are meant to pursue application goals, but the means to do so may change over time, either due to new opportunities or faults. Hence, also the wiring of event flows to funnels may need learning adaptations to perform to cope with the ever-changing environment typical of CAS.

10.7.3 Learn Deployment Policies

Another fundamental cornerstone of the Fluidware approach is *opportunistic deployment*, for achieving both scale-independence and robustness: the capability of deploying, migrating, replicating, and relocating processing components across the edge-to-cloud spectrum. Fluidware in fact does not foresee exclusively Edge deployments but welcomes a synergistic exploitation of all the tiers of a possible deployment: Edge to harness real-time processing on-board devices, Fog computing when battery life or computational power issues demand so, and Cloud when low latency is not a constraint or heavy analytics must be carried out. However, in this complex, multi-tiered deployment scenario, it could be difficult to define where to place each processing component at design time. Given the intrinsically dynamic and unpredictable nature of CAS, it would be much better to let the Fluidware middleware learn the best deployment configurations based on past experiences.

For instance, a Fluidware middleware may continuously monitor aspects such as battery life of devices, connectivity strength, processing time, data back pressure, as well as application requirements and the like to try to predict the need for migrating or relocating processing tasks. A device with a low battery level may stop processing data on-board and delegate to a Fog node, or in case the middleware predicts a rapid upsurge of incoming requests may decide to transfer processing to a Cloud node at the cost of latency.

All of this can be done without learning of course and is in fact the preferred way of doing these things nowadays. However, this puts lots of burden on system designer that must carefully engineer static policies to migrate and relocate components when needed. Having a system able to autonomously learn when and how to reconfigure deployment at run time would be incredibly valuable.

The more modular and fine-grained they are, the easier would be for applications to, for instance, learn how to compose them for achieving a desired goal.

10.8 Open Challenges

10.8.1 Multiagent Credit Assignment: How to Reward Agents for Their Contribution

A significant challenge in many-agent system is determining how to fairly reward individual agents for their contributions to collective goals, especially when their actions are interdependent. This is an open problem typically referred to as *multiagent credit assignment* in the literature. Even if several solutions have been proposed like difference rewards [47], ad hoc reward shaping [9], and ad hoc multiagent learning algorithms (like COMA [14]), the problem is still open in the general case, and it is even more challenging in the case of CAS, where agents are not necessarily homogeneous and very numerous. This could involve the development of new credit assignment algorithms that take into account the complex interactions between agents in such large-scale scenario.

10.8.2 Lightweight Learning: How to Learn with Limited Resources

In CAS environments computational resources are often severely limited. Devices like sensors, actuators, and edge nodes may have constraints in terms of processing power, memory, and energy availability. In such contexts, lightweight learning algorithms become not just beneficial but essential for real-time adaptability and efficient operation. Traditional machine learning algorithms often require significant computational resources for training and inference. This is impractical in IoT settings where devices are constrained by limited processing capabilities and power supplies. Therefore, there is a pressing need for algorithms that can operate under these constraints while still providing accurate and timely insights. Even if several solutions have been proposed like federated learning [21], they still do not consider the specificities of CAS, where there is a need for a collective outcome. Therefore, this could involve the development of new lightweight learning algorithms that take into account the specificities of CAS, considering both the resource constraints and the emergence of desired behaviors.

10.8.3 Lightweight Inference: How to Infer with Limited Resources

Similar to the challenges posed by lightweight learning, the need for efficient inference mechanisms is equally critical in CAS, particularly in edge–cloud computing scenarios. In these settings, data is often generated at the edge of the network, close to where it is needed most. However, the edge devices, such as IoT sensors and mobile phones, are typically constrained in terms of computational power, memory, and energy resources. As a result, the system must be capable of making quick and accurate inferences with these limited resources.

One approach to tackle this challenge is the use of specialized inference engines that are optimized for low-resource environments. These engines can be designed to perform specific types of calculations more efficiently, thereby reducing the computational load on the device. For example, TinyML is an emerging field that focuses on deploying machine learning algorithms on tiny, low-power hardware. Algorithms in this domain are optimized for quick inference and low energy consumption, making them ideal for edge computing scenarios.

Another strategy is the use of quantization techniques that simplify the numerical precision of the model parameters. By reducing the bit width of the model's weights and biases, the computational and memory requirements for inference can be significantly lowered. This comes at the cost of some loss in model accuracy, but in many real-world applications, this trade-off is acceptable.

Edge–cloud orchestration can also play a pivotal role in optimizing inference tasks. In scenarios where the edge device lacks the necessary computational power to perform the inference locally, the task can be offloaded to a nearby cloud server. The decision to offload can be made dynamically based on various factors such as the current load on the device, network latency, and the urgency of the inference task. Effective edge–cloud orchestration algorithms can help in making these decisions in real time.

Model pruning is another technique that can be employed to make the models more lightweight. By removing the neurons that contribute the least to the model's performance, the complexity of the neural network can be reduced, thereby speeding up the inference time.

10.9 Conclusion

Collective Adaptive Systems (CAS) offer a promising avenue for tackling complex, large-scale problems that are beyond the capabilities of individual agents. However, the dynamic and distributed nature of these systems introduces a host of challenges that need to be addressed for their effective deployment and operation. This chapter has provided an overview of the state of the art in learning techniques for CAS, focusing on the Fluidware framework as a case study. We have also identified several open challenges that represent important directions for future research.

References

1. Aguzzi, G., Casadei, R., Viroli, M.: Addressing collective computations efficiency: towards a platform-level reinforcement learning approach. In: Casadei, R., Nitto, E.D., Gerostathopoulos, I., Pianini, D., Dusparic, I., Wood, T., Nelson, P.R., Pournaras, E., Bencomo, N., Götz, S., Krupitzer, C., Raibulet, C. (eds.) IEEE International Conference on Autonomic Computing and Self-Organizing Systems, ACSOS 2022, Virtual, CA, USA, September 19–23, 2022, pp. 11–20. IEEE, Piscataway (2022). https://doi.org/10.1109/ACSOS55765.2022.00019
2. Aguzzi, G., Casadei, R., Viroli, M.: Machine learning for aggregate computing: a research roadmap. In: 42nd IEEE International Conference on Distributed Computing Systems, ICDCS Workshops, Bologna, Italy, July 10, 2022, pp. 119–124. IEEE, Piscataway (2022). https://doi.org/10.1109/ICDCSW56584.2022.00032
3. Aguzzi, G., Casadei, R., Viroli, M.: Towards reinforcement learning-based aggregate computing. In: ter Beek, M.H., Sirjani, M. (eds.) Coordination Models and Languages—24th IFIP WG 6.1 International Conference, COORDINATION 2022, Held as Part of the 17th International Federated Conference on Distributed Computing Techniques, DisCoTec 2022, Lucca, Italy, June 13–17, 2022, Proceedings. Lecture Notes in Computer Science, vol. 13271, pp. 72–91. Springer, Berlin (2022). https://doi.org/10.1007/978-3-031-08143-9_5
4. Aguzzi, G., Viroli, M., Esterle, L.: Field-informed reinforcement learning of collective tasks with graph neural networks. In: IEEE International Conference on Autonomic Computing and Self-Organizing Systems, ACSOS 2023, CA, Toronto, September 19–23, 2022 (2023)
5. Pianini, D., Casadei, R., Viroli, M., Mariani, S., Zambonelli, F.: Time-fluid field-based coordination through programmable distributed schedulers. Log. Methods Comput. Sci. **17**(4) (2021). https://doi.org/10.46298/lmcs-17(4:13)2021

6. Chou, Y., Moreira, C., Bruza, P., Ouyang, C., Jorge, J.A.: Counterfactuals and causability in explainable artificial intelligence: theory, algorithms, and applications. Inf. Fusion **81**, 59–83 (2022)
7. Christoffersen, P.J.K., Haupt, A.A., Hadfield-Menell, D.: Get it in writing: formal contracts mitigate social dilemmas in multi-agent RL. In: Agmon, N., An, B., Ricci, A., Yeoh, W. (eds.) Proceedings of the 2023 International Conference on Autonomous Agents and Multiagent Systems, AAMAS 2023, London, United Kingdom, 29 May 2023–2 June 2023, pp. 448–456. ACM, New York (2023). https://doi.org/10.5555/3545946.3598670
8. D'Angelo, M., Gerasimou, S., Ghahremani, S., Grohmann, J., Nunes, I., Pournaras, E., Tomforde, S.: On learning in collective self-adaptive systems: state of practice and a 3d framework. In: Litoiu, M., Clarke, S., Tei, K. (eds.) Proceedings of the 14th International Symposium on Software Engineering for Adaptive and Self-Managing Systems, SEAMS@ICSE 2019, Montreal, QC, Canada, May 25–31, 2019, pp. 13–24. ACM, New York (2019). https://doi.org/10.1109/SEAMS.2019.00012
9. Devlin, S., Kudenko, D.: Theoretical considerations of potential-based reward shaping for multi-agent systems. In: Sonenberg, L., Stone, P., Tumer, K., Yolum, P. (eds.) 10th International Conference on Autonomous Agents and Multiagent Systems (AAMAS 2011), Taipei, Taiwan, May 2–6, 2011, vol. 1–3, pp. 225–232. IFAAMAS (2011). http://portal.acm.org/citation.cfm?id=2030503&CFID=69153967&CFTOKEN=38069692
10. Dhokia, V., Essink, W.P., Flynn, J.M.: A generative multi-agent design methodology for additively manufactured parts inspired by termite nest building. CIRP Ann. **66**(1), 153–156
11. Dorigo, M., Stützle, T.: Ant colony optimization. MIT Press, Cambridge (2004)
12. Fadiga, K., Houzé, É., Diaconescu, A., Dessalles, J.: To do or not to do: finding causal relations in smart homes. In: IEEE International Conference on Autonomic Computing and Self-Organizing Systems (2021)
13. Foerster, J.N., Assael, Y.M., de Freitas, N., Whiteson, S.: Learning to communicate with deep multi-agent reinforcement learning. In: Lee, D.D., Sugiyama, M., von Luxburg, U., Guyon, I., Garnett, R. (eds.) Advances in Neural Information Processing Systems 29: Annual Conference on Neural Information Processing Systems 2016, December 5–10, 2016, Barcelona, Spain, pp. 2137–2145 (2016). https://proceedings.neurips.cc/paper/2016/hash/c7635bfd99248a2cdef8249ef7bfbef4-Abstract.html
14. Foerster, J.N., Farquhar, G., Afouras, T., Nardelli, N., Whiteson, S.: Counterfactual multi-agent policy gradients. In: McIlraith, S.A., Weinberger, K.Q. (eds.) Proceedings of the Thirty-Second AAAI Conference on Artificial Intelligence, (AAAI-18), the 30th innovative Applications of Artificial Intelligence (IAAI-18), and the 8th AAAI Symposium on Educational Advances in Artificial Intelligence (EAAI-18), New Orleans, Louisiana, USA, February 2–7, 2018, pp. 2974–2982. AAAI Press (2018). https://doi.org/10.1609/aaai.v32i1.11794
15. Gronauer, S., Diepold, K.: Multi-agent deep reinforcement learning: a survey. Artif. Intell. Rev. **55**(2), 895–943 (2022). https://doi.org/10.1007/s10462-021-09996-w
16. Guo, R., Cheng, L., Li, J., Hahn, P.R., Liu, H.: A survey of learning causality with data: Problems and methods. ACM Comput. Surv. **53**(4), 1–37 (2020)
17. Heinze-Deml, C., Maathuis, M.H., Meinshausen, N.: Causal structure learning. Ann. Rev. Stat. Appl. **5**(1), 371–391 (2018)
18. Hernandez-Leal, P., Kartal, B., Taylor, M.E.: A very condensed survey and critique of multiagent deep reinforcement learning. In: Seghrouchni, A.E.F., Sukthankar, G., An, B., Yorke-Smith, N. (eds.) Proceedings of the 19th International Conference on Autonomous Agents and Multiagent Systems, AAMAS '20, Auckland, New Zealand, May 9–13, 2020, pp. 2146–2148. International Foundation for Autonomous Agents and Multiagent Systems (2020). https://doi.org/10.5555/3398761.3399105
19. Heylighen, F.: Stigmergy as a generic mechanism for coordination: definition, varieties and aspects. Vrije Universiteit Brussels, ECCO working paper edn. (2012)
20. Izquierdo-Torres, E.: Collective intelligence in multi-agent robotics: stigmergy, self-organization and evolution. Brighton BNI (2004)

21. Khan, L.U., Saad, W., Han, Z., Hossain, E., Hong, C.S.: Federated learning for Internet of Things: recent advances, taxonomy, and open challenges. IEEE Commun. Surv. Tutorials **23**(3), 1759–1799 (2021). https://doi.org/10.1109/COMST.2021.3090430
22. Lalanda, P., Hamon, C., Escoffier, C., Lévèque, T.: iCasa, a development and simulation environment for pervasive home applications. In: IEEE 11th Consumer Communications and Networking Conference (2014)
23. Leibo, J.Z., Hughes, E., Lanctot, M., Graepel, T.: Autocurricula and the emergence of innovation from social interaction: a manifesto for multi-agent intelligence research. CoRR abs/1903.00742 (2019). http://arxiv.org/abs/1903.00742
24. Lippi, M., Mariani, S., Martinelli, M., Zambonelli, F.: Individual and collective self-development: concepts and challenges. In: Proceedings of the 17th Conference on Computer Science and Intelligence Systems, FedCSIS, Sofia, Bulgaria, September 4–7, 2022
25. Lippi, M., Mariani, S., Zambonelli, F.: Developing a "sense of agency" in IoT systems: preliminary experiments in a smart home scenario. In: 19th IEEE International Conference on Pervasive Computing and Communications Workshops (2021)
26. Mahfoudh, H.B., Serugendo, G.D.M., Naja, N., Abdennadher, N.: Learning-based coordination model for spontaneous self-composition of reliable services in a distributed system. Int. J. Softw. Tools Technol. Transf. **22**(4), 417–436 (2020). https://doi.org/10.1007/s10009-020-00557-0
27. Mamei, M., Menezes, R., Tolksdorf, R., Zambonelli, F.: Case studies for self-organization in computer science. J. Syst. Archit. **52**(8–9), 443–460 (2006)
28. Mariani, S., Roseti, P., Zambonelli, F.: Multi-agent learning of causal networks in the Internet of Things. In: Mathieu, P., Dignum, F., Novais, P., De la Prieta, F. (eds.) Advances in Practical Applications of Agents, Multi-Agent Systems, and Cognitive Mimetics. The PAAMS Collection, pp. 163–174. Springer, Cham (2023)
29. Mariani, S., Zambonelli, F.: Learning stigmergic communication for self-organising coordination. In: IEEE International Conference on Autonomic Computing and Self-Organizing Systems, ACSOS 2023, Toronto, ON, Canada, September 25–29, 2023, pp. 47–56. IEEE, Piscataway (2023). https://doi.org/10.1109/ACSOS58161.2023.00022
30. Oroojlooyjadid, A., Hajinezhad, D.: A review of cooperative multi-agent deep reinforcement learning (2019). CoRR abs/1908.03963. http://arxiv.org/abs/1908.03963
31. Panait, L., Luke, S.: Cooperative multi-agent learning: the state of the art. Auton. Agents Multi Agent Syst. **11**(3), 387–434 (2005). https://doi.org/10.1007/s10458-005-2631-2
32. Pearl, J.: The seven tools of causal inference, with reflections on machine learning. Commun. ACM **62**(3), 54–60 (2019)
33. Pearl, J., Mackenzie, D.: The Book of Why. Basic Books, New York (2018)
34. Pinciroli, C., Lee-Brown, A., Beltrame, G.: Buzz: an extensible programming language for self-organizing heterogeneous robot swarms (2015). CoRR abs/1507.05946. http://arxiv.org/abs/1507.05946
35. Pollack, J., Bedau, M.A., Husbands, P., Watson, R.A., Ikegami, T.: Learning Ant Foraging Behaviors (2004)
36. Razian, M.R., Fathian, M., Bahsoon, R., Toosi, A.N., Buyya, R.: Service composition in dynamic environments: a systematic review and future directions. J. Syst. Softw. **188**, 111290 (2022). https://doi.org/10.1016/j.jss.2022.111290
37. Reid, C.R., Latty, T.: Collective behaviour and swarm intelligence in slime moulds. FEMS Microbiol. Rev. **40**(6), 798–806 (2016)
38. Scherrer, N., Bilaniuk, O., Annadani, Y., Goyal, A., Schwab, P., Schölkopf, B., Mozer, M.C., Bengio, Y., Bauer, S., Ke, N.R.: Learning neural causal models with active interventions (2021). CoRR abs/2109.02429. https://arxiv.org/abs/2109.02429
39. Schmickl, T., Crailsheim, K.: A navigation algorithm for swarm robotics inspired by slime mold aggregation. In: Swarm Robotics, Second International Workshop, Rome, Italy, September 30–October 1, 2006
40. Schölkopf, B., Locatello, F., Bauer, S., Ke, N.R., Kalchbrenner, N., Goyal, A., Bengio, Y.: Toward causal representation learning. Proc. IEEE **109**(5), 612–634 (2021)

41. Solar-Lezama, A.: Program Synthesis by Sketching. Ph.D. Thesis, USA (2008), aAI3353225
42. Sosic, A., KhudaBukhsh, W.R., Zoubir, A.M., Koeppl, H.: Inverse reinforcement learning in swarm systems. In: Larson, K., Winikoff, M., Das, S., Durfee, E.H. (eds.) Proceedings of the 16th Conference on Autonomous Agents and MultiAgent Systems, AAMAS 2017, São Paulo, Brazil, May 8–12, 2017, pp. 1413–1421. ACM, New York (2017). http://dl.acm.org/citation.cfm?id=3091320
43. Stone, P., Veloso, M.M.: Multiagent systems: a survey from a machine learning perspective. Auton. Robots **8**(3), 345–383 (2000). https://doi.org/10.1023/A:1008942012299
44. Sunder, V., Vig, L., Chatterjee, A., Shroff, G.: Prosocial or selfish? Agents with different behaviors for contract negotiation using reinforcement learning (2018). CoRR abs/1809.07066. http://arxiv.org/abs/1809.07066
45. Viroli, M., Beal, J., Damiani, F., Audrito, G., Casadei, R., Pianini, D.: From field-based coordination to aggregate computing. In: Serugendo, G.D.M., Loreti, M. (eds.) Coordination Models and Languages—20th IFIP WG 6.1 International Conference, COORDINATION 2018, Held as Part of the 13th International Federated Conference on Distributed Computing Techniques, DisCoTec 2018, Madrid, Spain, June 18–21, 2018. Proceedings. Lecture Notes in Computer Science, vol. 10852, pp. 252–279. Springer, Berlin (2018). https://doi.org/10.1007/978-3-319-92408-3_12
46. Vowels, M.J., Camgöz, N.C., Bowden, R.: D'ya like DAGs? A survey on structure learning and causal discovery (2021). CoRR abs/2103.02582 (2021). https://arxiv.org/abs/2103.02582
47. Wolpert, D.H., Tumer, K.: Optimal payoff functions for members of collectives. Adv. Complex Syst. **4**(2–3), 265–280 (2001). https://doi.org/10.1142/S0219525901000188
48. Ye, D., Zhang, M., Vasilakos, A.V.: A survey of self-organization mechanisms in multiagent systems. IEEE Trans. Syst. Man Cybern. Syst. **47**(3), 441–461 (2017). https://doi.org/10.1109/TSMC.2015.2504350
49. Zaïem, M.S., Bennequin, E.: Learning to communicate in multi-agent reinforcement learning: a review (2019). CoRR abs/1911.05438. http://arxiv.org/abs/1911.05438
50. Zambonelli, F., Omicini, A., Anzengruber, B., Castelli, G., Angelis, F.L.D., Serugendo, G.D.M., Dobson, S.A., Fernandez-Marquez, J.L., Ferscha, A., Mamei, M., Mariani, S., Molesini, A., Montagna, S., Nieminen, J., Pianini, D., Risoldi, M., Rosi, A., Stevenson, G., Viroli, M., Ye, J.: Developing pervasive multi-agent systems with nature-inspired coordination. Pervasive Mobile Comput. **17**, 236–252 (2015)
51. Zhang, K., Yang, Z., Basar, T.: Multi-agent reinforcement learning: a selective overview of theories and algorithms (2019). CoRR abs/1911.10635. http://arxiv.org/abs/1911.10635
52. Zhang, K., Yang, Z., Basar, T.: Decentralized multi-agent reinforcement learning with networked agents: recent advances. Frontiers Inf. Technol. Electron. Eng. **22**(6), 802–814 (2021). https://doi.org/10.1631/FITEE.1900661
53. Zhu, C., Dastani, M., Wang, S.: A survey of multi-agent reinforcement learning with communication (2022). CoRR abs/2203.08975. https://doi.org/10.48550/arXiv.2203.08975

Index

F. Zambonelli et al. (eds.), *Fluidware*, Internet of Things,
https://doi.org/10.1007/978-3-031-62146-8

SPRINGER NATURE

GPSR Compliance

The European Union's (EU) General Product Safety Regulation (GPSR) is a set of rules that requires consumer products to be safe and our obligations to ensure this.

If you have any concerns about our products, you can contact us on ProductSafety@springernature.com

In case Publisher is established outside the EU, the EU authorized representative is:

Springer Nature Customer Service Center GmbH
Europaplatz 3
69115 Heidelberg, Germany

The manufacturer's authorised representative in the EU is Springer Nature Customer Service Centre GmbH, Europaplatz 3, 69115 Heidelberg, Germany. If you have any concerns regarding our products, please contact ProductSafety@springernature.com

Printed and bound by CPI Group (UK) Ltd, Croydon, CR0 4YY
30/09/2026
02225584-0001